ELECTROANALYTICAL CHEMISTRY
A Series of Advances

VOLUME 26

ELECTROANALYTICAL CHEMISTRY
A Series of Advances

VOLUME 26

edited by

Allen J. Bard
and
Cynthia G. Zoski

CRC Press
Taylor & Francis Group
Boca Raton London New York

CRC Press is an imprint of the
Taylor & Francis Group, an **informa** business

CRC Press
Taylor & Francis Group
6000 Broken Sound Parkway NW, Suite 300
Boca Raton, FL 33487-2742

First issued in paperback 2019

© 2016 by Taylor & Francis Group, LLC
CRC Press is an imprint of Taylor & Francis Group, an Informa business

No claim to original U.S. Government works

ISBN-13: 978-1-4987-3377-9 (hbk)
ISBN-13: 978-0-367-37706-9 (pbk)

Visit the Taylor & Francis Web site at
http://www.taylorandfrancis.com

and the CRC Press Web site at
http://www.crcpress.com

Contents

Introduction to the Series

This series is designed to provide authoritative reviews in the field of modern electroanalytical chemistry defined in its broadest sense. Coverage is comprehensive and critical. Enough space is devoted to each chapter of each volume to include derivations of fundamental equations, detailed descriptions of apparatus and techniques, and complete discussion of important articles, so that the chapters may be read without repeated reference to the periodical literature. Chapters vary in length and subject area. Some are reviews of recent developments and applications of well-established techniques, whereas others contain discussion of the background and problems in areas still being investigated extensively and in which many statements may still be tentative. Finally, chapters on techniques generally outside the scope of electroanalytical chemistry, but which can be applied fruitfully to electrochemical problems, are included.

Electroanalytical chemists and others are concerned not only with the application of new and classical techniques to analytical problems, but also with the fundamental theoretical principles upon which these techniques are based.

Electroanalytical techniques are proving useful in such diverse fields as electroorganic synthesis, fuel cell studies, and radical ion formation as well as with such problems as the kinetics and mechanisms of electrode reactions, and the effects of electrode surface phenomena, adsorption, and the electrical double layer on electrode reactions.

It is hoped that the series is proving useful to the specialist and nonspecialist alike—that it provides a background and a starting point for graduate students undertaking research in the areas mentioned, and that it also proves valuable to practicing analytical chemists interested in learning about and applying electroanalytical techniques. Furthermore, electrochemists and industrial chemists with problems of electrosynthesis, electroplating, corrosion, and fuel cells, as well as other chemists wishing to apply electrochemical techniques to chemical problems, may find useful material in these volumes.

Allen J. Bard
Cynthia G. Zoski

Contributors to Volume 26

Shigeru Amemiya
Department of Chemistry
University of Pittsburgh
Pittsburgh, Pennsylvania

Lane A. Baker
Department of Chemistry
Indiana University
Bloomington, Indiana

Avni Berisha
Chemistry Department of Natural
    Sciences Faculty
University of Prishtina
Prishtina, Kosovo

Mohamed M. Chehimi
ICMPE, SPC, PoPI team, UPEC
Université Paris-Est
Thiais, France

Jean Pinson
ITODYS
Université Paris Diderot
Paris, France

Fetah I. Podvorica
Chemistry Department of Natural
    Sciences Faculty
University of Prishtina
Prishtina, Kosovo

Wenqing Shi
Department of Chemistry
Indiana University
Bloomington, Indiana

Anna Weber
Department of Chemistry
Indiana University
Bloomington, Indiana

Contents of Other Series Volumes

1 Nanoscale Scanning Electrochemical Microscopy

Shigeru Amemiya

CONTENTS

1.1   INTRODUCTION

Enabling scanning electrochemical microscopy (SECM) at the nanoscale (1–100 nm) has been the ultimate goal in the developmental of this powerful electrochemical method since its invention. In late 1980s [1–3], SECM was invented by Bard and coworkers as the unique scanning probe microscopy technique that employs a small electrode as a probe to quantitatively image and measure dynamic chemical processes at and near the interfaces. Immediately after the invention, they also pioneered studies on nanoscale SECM in early 1990s [4–6], which were followed by impactful but rather intermittent progresses until recently because of various nanotechnological challenges. In the meantime, micrometer-scale SECM was quickly established, widely adapted, and then commercialized by late 1990s to find a wide range of applications from electrochemistry to various fields of chemistry, biology, and materials science [7]. The success of micrometer-scale SECM is due not only to its technological simplicity and maturity but also to the versatility of the SECM principle, which allows for the study of various types of interfaces including solid/liquid, liquid/liquid, and gas/liquid interfaces, as well as biological membranes. However, present paradigm shifts in science and technology toward the nanoscale demand the transformation of SECM from micrometer scale to nanometer scale to further expand its utility for nanoscience and nanotechnology. There are urgent needs to image and quantitatively characterize the chemical reactivity of single nano-entities [8], thereby augmenting the significance of nanoscale SECM. Micrometer-scale SECM has been successfully applied to study the ensembles of nanoscale objects but rarely individual nanostructures [9]. Importantly, the consideration of nanometer dimensions in SECM theories has revealed that the rewards of nanoscale SECM go far beyond a better spatial resolution. The power of SECM will be fully realized only when it is operated at the nanoscale.

While many reviews [10–11] and a couple of comprehensive monographs [7,12] about SECM have been already published, this chapter is uniquely and timely devoted to nanoscale SECM, which has seen major progresses in the past several years. Specifically, we will overview the fundamental development of nanoscale SECM in four areas as the foundation of its future applications. The first area is focused on nanoscale imaging, which is achieved by employing a nanoelectrode as an SECM tip. Remarkably, spatial resolution of 1–100 nm, which is defined in this chapter as truly nanoscale [13] in contrast to submicrometer scale of 100–1000 nm, has been demonstrated to image individual nanoscale entities in biological and nanomaterial systems as substrates. The second area is concerned about the SECM measurement based on a nanometer-wide gap between the tip and the substrate at fixed positions. The nanogap-based approach enables the electrochemical detection of single molecules and the kinetic study of extremely fast chemical processes at the interfaces and within the nanogap. The progresses in these two areas have been made by the development or introduction of several enabling technologies. This third area covers various nanotechnologies not only for the fabrication and characterization of a nanotip as anticipated challenges but also for nanotip handling and positioning as the manifestation of adventitious environmental effects on nanoscale electrochemical measurement. Finally, a nanoelectrode can be functionalized to combine SECM with other imaging techniques for multidimensional nanoscale imaging. Highly sophisticated multifunctional nanotips have been developed to interconnect SECM with atomic force microscopy (AFM), scanning ion conductance microscopy (SICM), and spectroelectrochemistry at the nanoscale. Developmental efforts on nanoscale SECM in these four areas are complementary to recent efforts on the applications of micrometer-scale SECM to a variety of real nanosystems with fundamental or practical importance [9]. A future combination between these two efforts will synergistically open up exciting new applications of nanoscale SECM as briefly discussed in the last section of this chapter.

1.2 NANOSCALE IMAGING

Nanoscale SECM imaging requires the tip that is smaller than 1 μm to achieve nanometer-scale lateral resolution. Typically, submicrometer-sized tips (100–1000 nm) have been employed, although truly nanoscale resolution of 1–100 nm [13] was also reported. In this chapter, SECM imaging with a nanotip is considered as nanoscale imaging even when the tip is laterally moved with micrometer resolution because the nanotip must be scanned at a nanoscale distance from a substrate to observe a feedback effect (see below). In fact, nanoscale SECM imaging is mainly performed with the feedback mode, which has been well established for micrometer-scale SECM since its invention [2,3,14]. In this operation mode, a tip is positioned within a tip diameter from a substrate during imaging to study a feedback effect from the substrate on tip current as a measure of substrate reactivity. The resultant lateral resolution is determined by tip size and tip–substrate distance. In contrast, the lateral resolution of the substrate generation/tip collection (SG/TC) mode is limited by substrate size, when a nanometer-sized tip is scanned further than a tip diameter from the substrate to monitor the concentration profile of a substrate-generated species [15–16]. Importantly, the spatial resolution of the SG/TC mode is equivalent to that of the feedback counterpart when SECM is operated with a nanometer tip–substrate distance [17].

1.2.1 RESOLUTION AND DETECTABILITY

The lateral resolution of SECM is defined by the ability to resolve two nearby objects and is different from the ability to detect an isolated small object, that is, detectability. The lateral resolution and detectability of SECM in the feedback mode were assessed theoretically, when a disk-shaped tip was positioned over the disk-shaped conductive spot embedded in an insulating substrate (Figure 1.1a) [18]. In the feedback mode, a redox mediator in solution, O, is electrolyzed amperometrically at the tip ($O + e \rightarrow R$) to yield steady-state diffusion-limited current. When the tip is positioned within a tip diameter from the conductive spot, tip current is enhanced

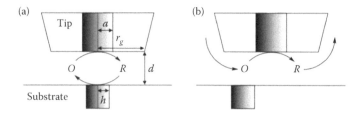

FIGURE 1.1 (a) Positive and (b) negative feedback effects on a disk-shaped tip from a substrate with the small active spot embedded in an insulator.

by a feedback effect from the active spot, which regenerates the original mediator $(R \rightarrow O + e)$ to be detected at the tip. The resultant feedback current is higher at a shorter tip–spot distance and can be as high as expected with an infinitely large conductive substrate when the radius of the active spot, h, is larger than a threshold value, h^∞, as given by Reference [18]

$$h^\infty = a + 1.5d \tag{1.1}$$

where a is tip radius and d is tip–substrate distance (Figure 1.1a). The corresponding diameter, $2h^\infty$, defines the lateral resolution of SECM, which is the size of the local substrate surface that is actually "seen" in the SECM feedback experiment. Equation 1.1 indicates that h^∞ is eventually limited by a as d approaches zero. Practically, d is larger than $0.1a$ owing to tip–substrate misalignment (Section 1.3.2.1), thereby compromising lateral resolution to $>2.3a$. Noticeably, the electroactive spot that is comparable to or smaller than the tip size must be externally biased to maintain steady-state mediator regeneration $(R \rightarrow O + e)$ [19–20].

Importantly, SECM can detect the local active spot that is smaller than the tip size, where tip current is still higher than expected over an insulating substrate (Figure 1.1b). When the tip is positioned closer to the insulating surface, tip current is lowered by the substrate, which only hinders diffusion of a redox mediator to the tip without its regeneration. Since the insulating sheath of the tip also hinders mediator diffusion, the resultant negative feedback effect on tip current depends not only on the tip–substrate distance but also on the outer radius of the insulating sheath, r_g, which is often given by $RG(=r_g/a)$. Overall, the difference of tip current with and without the active spot under a tip increases at a shorter tip–substrate distance. With a short tip–substrate distance of $0.1a$, the enhancement of tip current is detectable when the active spot is 10–20 times smaller than the tip diameter [18]. For instance, a 5- to 10-nm-diameter active spot can be detected by using the 100-nm-diameter tip positioned at 5 nm from the substrate. In this case, the spatial resolution is still determined by the corresponding $2h^\infty$ value of 115 nm (Equation 1.1) so that the 5- to 10-nm-diameter spot appears as ~115-nm-diameter spot in the SECM image. Thus, the shape of the active spot is not resolvable directly from the SECM image while the tip current over the small active spot reflects its size.

The detectability of SECM is much higher for an electroactive spot with a high aspect ratio as demonstrated by using a nanoband electrode as a model substrate [21–23].

For instance, ultra long single-walled carbon nanotubes with a diameter of only ~1.6 nm were grown on the SiO_2-coated Si wafer to detect individual nanotubes by using 1.5- to 10-μm-diameter SECM tips in the feedback mode [24]. This superb detectability is ascribed to the efficient nanoscale mass transport of a redox mediator at the nanotubes. In addition, the nanotubes were much longer than the tip diameter to give a substantial feedback response at steady states under unbiased conditions.

1.2.2 CONSTANT-HEIGHT AND CONSTANT-DISTANCE IMAGING

Nanoscale SECM imaging typically employs the constant-height or constant-distance mode to scan a tip within a feedback distance from a substrate. In the constant-height mode, the vertical position of a tip is unchanged during its lateral scan over a substrate. This simple operation mode was the first to enable nanoscale SECM imaging [4]. In this pioneering work, a Pt tip with an effective diameter of down to ~0.2 μm was developed to image an interdigitated array electrode. The amperometric tip current based on the diffusion-limited reduction of methyl viologen was monitored during constant-height imaging (Figure 1.2a). A contrast between high and low tip currents in the resultant image is due to positive and negative feedback effects on tip current from the conductive (Pt) and insulating (SiO_2) portions of the substrate, respectively. In general, a constant-height image is the convolution of the reactivity and topography of the substrate surface because tip current depends on the local reactivity of the substrate surface under the tip and the corresponding local tip–substrate distance. In Figure 1.2a, surface reactivity dominated the contrast of the constant-height image of the relatively flat array electrode, whereas topography information such as a difference in height between Pt and SiO_2 surfaces was not obtained. In fact, the surface of the array electrode was as flat as needed to maintain

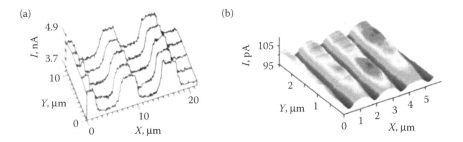

FIGURE 1.2 (a) Constant-height image of an interdigitated array electrode obtained by using a glass-sealed Pt tip (0.2 μm in effective diameter) in 2 M KCl solution of 0.2 M methyl viologen as a redox mediator. (Reprinted with permission from Lee C., C. J. Miller, and A. J. Bard, Scanning electrochemical microscopy: Preparation of submicrometer electrodes, *Anal. Chem.*, Vol. 63, 1991: pp. 78–83. Copyright 1991, American Chemical Society.) (b) Constant-height image of a gold CD surface obtained with a 140-nm-radius Pt tip. Tip scan rate, 300 nm/s. The solution contained 1 mM $FcCH_2OH$ and 250 mM KCl and was buffered with acetic acid to pH = 4. (Reprinted with permission from Laforge, F. O. et al., Nanoscale imaging of surface topography and reactivity with the scanning electrochemical microscope, *Anal. Chem.*, Vol. 81, 2009: pp. 3143–3150. Copyright 2009, American Chemical Society.)

a feedback distance between the 0.2-μm-diameter tip and the substrate without their contact. The surface of most substrates, however, is rough at the nanometer scale, thereby limiting the utility of the constant-height mode for nanoscale imaging.

A recent study also demonstrated a nanoscale lateral resolution by employing the constant-height mode [25]. An important aspect of this work is the usage of reliable Pt nanotips with well-characterized shape and size (Section 1.4.2.2). Figure 1.2b shows the constant-height image of a gold-coated compact disk as obtained by using an inlaid disk Pt tip with a radius of 140 nm. The high-resolution image shows a few "ridges" and "valleys" within the small 2.5 μm × 2.5 μm portion of the gold-coated surface. The respective features with higher and lower tip currents are mainly due to surface topography. The ridges are higher and closer to the tip to give more positive tip current based on the oxidation of ferrocenemethanol (FcMeOH), whereas lower tip current over the valleys is due to a longer distance from the tip. Interestingly, significant local variations in tip current over each ridge demonstrate the nonuniform reactivity of the ridge surface, which is very flat as confirmed by AFM. The surface reactivity, however, cannot be quantitatively determined from the constant-height image without the knowledge of surface topography under the tip. Therefore, the so-called approach curve was measured to determine surface reactivity separately. Specifically, tip current was monitored as the tip approached vertically to the ridge, which was located by imaging. A plot of tip current versus tip–substrate distance, that is, approach curve, was fitted with a theoretical curve to determine a heterogeneous electron-transfer rate constant of 0.50 cm/s.

The limitations of the constant-height mode can be overcome by employing the constant-distance mode. In this imaging mode, the vertical tip position is adjusted during lateral tip scan to maintain a constant distance between the tip and the substrate. Thus, a plot of tip current versus x-, y-positions represents the image of substrate reactivity, while the topography image of the substrate is obtained from changes in vertical tip position. Constant-distance imaging, however, is straightforward only at a rough substrate with uniform reactivity or at a flat substrate with heterogeneous reactivity. In the latter case, constant-distance imaging is equivalent to constant-height imaging. In the former case, a topography image can be obtained by employing the constant-current mode, where the vertical tip position is adjusted during imaging to maintain a constant tip current and, subsequently, a constant tip–substrate distance. The precision of maintaining a constant tip–substrate distance is eventually limited by the high vertical resolution of SECM as estimated to be ~0.02a, for example, 1 nm with a 100-nm-diameter tip [26]. When the substrate surface is heterogeneously reactive and rough, more sophisticated protocols are required to separately obtain reactivity and topography images by using a nanoelectrode tip (Section 1.2.3). Alternatively, a multifunctional tip is required to separately monitor the reactivity and topography of the complicated substrate as discussed in Section 1.5.

Nanoscale constant-current imaging was demonstrated for non-flat substrates with relatively uniform reactivity [25]. Figure 1.3a shows the constant-current SECM image of the 57 μm × 57 μm portion of the EPROM of a Motorola 68HC05 chip obtained with a 100-nm-radius Pt tip. The tip current based on FcMeOH oxidation was maintained

(a) (b)

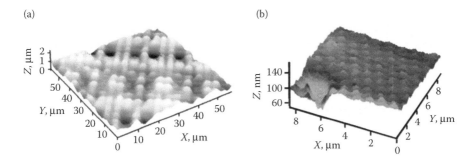

FIGURE 1.3 (a) Constant-current image of the 57 μm × 57 μm area of the EPROM of the Motorola 68HC05 chip in 1 mM FcMeOH aqueous solution as obtained with a 100-nm-radius Pt tip. The tip/substrate distance was ~200 nm, and the current was set at 21 pA (70% of the bulk value). The imaging time was 20 min at the 3 μm/s scan rate. (b) Constant-current image of an IBM wafer showing a microscopic defect as obtained with a 190-nm-radius Pt tip. The solution contained 5 mM $Ru(NH_3)_6Cl_3$ and 250 mM KCl. Tip scan rate, 500 nm/s. (Reprinted with permission from Laforge, F. O. et al., Nanoscale imaging of surface topography and reactivity with the scanning electrochemical microscope, *Anal. Chem.*, Vol. 81, 2009: pp. 3143–3150. Copyright 2009, American Chemical Society.)

at 70% of the bulk value in the negative feedback mode to hold the tip at 200 nm away from the insulating substrate during imaging. The topographic image exhibits a maximum height difference of ~2 μm, which corresponds to ~20 times of the tip radius. Such a non-flat substrate cannot be imaged in the constant-height mode without crashing the tip into the protruded regions of the substrate or scanning the tip far away from the recessed regions of the substrate. The constant-current mode was also employed to find an active spot as a large "dip" in the 9 μm × 8 μm image of the IBM wafer built with the 90 nm process technology (Figure 1.3b). The high-resolution image was obtained by amperometrically monitoring $Ru(NH_3)_6Cl_3$ reduction at a 190-nm-radius Pt tip. The dip is not due to substrate topography and was not seen in an optical micrograph. The dip reflects the enhancement of tip current by regeneration of the redox mediator at the active spot surrounded by the original insulating surface with a square array of sub-micrometer-sized, pyramidal bumps. The reactivity of the dip region was still so low that the tip was moved closer to the dip-like feature to maintain the pre-set current value that was lower than the tip limiting current in the bulk solution, thereby enabling constant-current imaging of the mainly insulating surface. The local surface reactivity, however, cannot be determined quantitatively from tip current without somehow knowing the topography of the dip region of the substrate. Moreover, the constant-current mode is not suitable for imaging the highly active spot of an insulating substrate because a tip over the respective surfaces gives positive and negative feedback currents [27].

1.2.3 REACTIVITY AND TOPOGRAPHY IMAGING

The nanoscale topography and reactivity images of a heterogeneously reactive and non-flat substrate can be obtained separately and simultaneously by amperometrically detecting two probe molecules or by measuring tip–substrate interactions in addition to tip current. For instance, the former approach was enabled by employing a nanopipette-supported interface between two immiscible electrolyte solutions (ITIES) as an SECM tip [28]. Specifically, a 270-nm-radius glass nanopipette was filled with the 1,2-dichloroethane (DCE) solution of ferro-cenedimethanol (FDM) as a probe molecule for reactivity imaging. The DCE-filled nanopipette was immersed in the aqueous solution of PF_6^- as the second probe molecule for topography imaging. A reactivity image was obtained by studying the current based on FDM oxidation at the substrate. This redox-active mediator can partition from the inner DCE solution into the adjacent external aqueous solution to be oxidized at a 25-μm-diameter Pt disk electrode as a substrate (Figure 1.4a), thereby yielding high substrate current (Figure 1.4b). In contrast, substrate current was much lower when the tip was moved over the inert glass surface surrounding

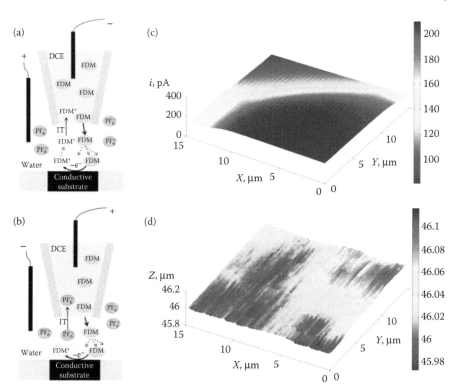

FIGURE 1.4 Schematic representation of (a) positive and (b) negative ion-transfer feedback modes. The respective operation modes were employed to obtain (c) reactivity and (d) topography images of the 12.5-μm-radius Pt disk embedded in glass by using a 270-nm-radius nanopipette tip. The scan rate was 500 nm/s. The nanopipette was filled with the DCE solution of 26 mM FDM. External aqueous solution contained 0.46 mM $LiPF_6$. (Wang, Y. et al., Electron transfer/ion transfer mode of scanning electrochemical microscopy (SECM): A new tool for imaging and kinetic studies, *Chem. Sci.*, Vol. 4, 2013: pp. 3606–3616. Reproduced by permission of the Royal Society of Chemistry.)

the Pt disk. This reactivity image also detailed the geometry of the Pt/glass boundary including small protrusions of Pt into glass as found in the optical micrograph. Importantly, the reactivity image based on substrate current was not affected by substrate topography. A constant tip–substrate distance was maintained during imaging by employing the constant-current mode based on the transfer of PF_6^- at the nanopipette-supported ITIES tip (Figure 1.4c). Since PF_6^- is not electroactive, constant tip current in the negative feedback mode was set to maintain a constant tip–substrate distance (Figure 1.4d). This topography image shows that the well-polished substrate including the Pt/glass boundary is flat at the nanoscale and is slightly tilt as indicated by ~100-nm variation between the lower left and the upper right corners of the image (~20 μm distance).

More recently, the voltage-switching mode was proposed for simultaneous topography and reactivity imaging of single biological cells by switching tip potentials for the detection of two probe molecules [29]. Specifically, the voltage-switching mode was applied for the constant-distance imaging of single A431 cells to determine the distribution of epidermal growth factor receptors. The receptors were labeled with alkaline-phosphatase-tagged antibody for reactivity imaging. A constant tip–cell distance was maintained by employing $Ru(NH_3)_6^{3+}$ as a membrane-impermeable probe to obtain a topographic image in the negative feedback mode. Noticeably, the voltage-switching mode was combined with the hopping method to trace the rough surface of the cell. This tapping-mode-like approach was originally introduced by Bard and coworkers for simultaneous topography, electrochemical, and optical imaging [30]. Figure 1.5a illustrates the tip movement and the timing of current measurement during the imaging process. Initially, the potential of ~0.72-μm-radius carbon tip was set at −0.5 V (vs. Ag/AgCl) for reduction of $Ru(NH_3)_6^{3+}$. The tip approached toward the cell surface until the tip current decreased to a preset value due to a negative feedback effect from the cell surface. As soon as the preset current was achieved to assure a constant tip–cell distance, the tip potential was switched to +0.35 V to detect p-aminophenol in the SG/TC mode. This probe molecule was generated from p-aminophenyl phosphate by alkaline phosphatase to locate epidermal growth factor receptors. Then, the tip was withdrawn for a preset distance and moved to the next lateral position, where the negative feedback and SG/TC measurements were repeated at the constant tip–cell distance. Overall, a topography image was obtained from the vertical tip position at the end of each tip approach, while the tip current based on p-aminophenol oxidation at the final tip position gave an electrochemical image (Figure 1.5b). The constant-distance electrochemical image demonstrates the heterogeneous distribution of epidermal growth factor receptors, although the spatial resolution was limited by the relatively large size of the carbon-filled micropipette tip. This tip, however, was small enough to quickly yield a steady-state current response upon potential switch. This quick response is supported also by the theory of SECM chronoamperometry, which predicts that a steady-state current response is obtained within a time of $10a^2/D$ at $d = a$ from an inert substrate [31], that is, within 5 ms with $a = 0.72$ μm and $D = 1 \times 10^{-5}$ cm^2/s.

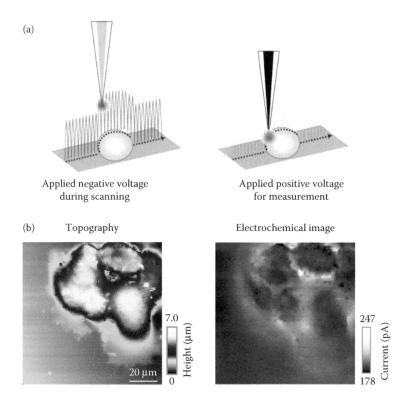

(a)

Applied negative voltage   Applied positive voltage
during scanning   for measurement

(b)   Topography   Electrochemical image

7.0
Height (μm)
20 μm
0

247
Current (pA)
178

FIGURE 1.5 (a) Voltage-switching mode of SECM for (left) topography and (right) reactivity imaging. The carbon tip was moved by using the hopping method. (b) Topography (left) and electrochemical (right) images of A431 cells in a HEPES buffer containing 10 mM $Ru(NH_3)_6Cl_3$ and 4.7 mM p-aminophenyl phosphate. (Reprinted with permission from Takahashi, Y. et al., *Proc. Natl. Acad. Sci. USA*, Vol. 109, 2012: pp. 11540–11545.)

The intermittent-contact mode uses interactions between the tip and the substrate for topography imaging, while tip current is simultaneously monitored for reactivity imaging [32]. In this imaging mode, a nanotip was attached to the piezoelectric bender actuator that oscillated perpendicularly to a substrate (Figure 1.6a). A tip–substrate contact was detected as the dumping of the oscillation of piezoelectric bender by using its servomechanism, which returned a sensor signal as a measure of the actual oscillation amplitude. This sensing mechanism differentiates the intermittent-contact mode from the tip-position modulation mode [27], which monitors tip current to measure the amplitude of tip oscillation. Advantageously, a piezoelectric positioner with servo control can be readily implemented into an SECM instrument to simplify the experimental setup in comparison to the other distance-control methods based on the measurement of tip–substrate interactions, that is, SECM–AFM and the nanoscale shear-force [33–35] mode (Sections 1.5.2 and 1.5.4, respectively). The general limitation of these contact-imaging methods, however, is the potential damage of the tip and the

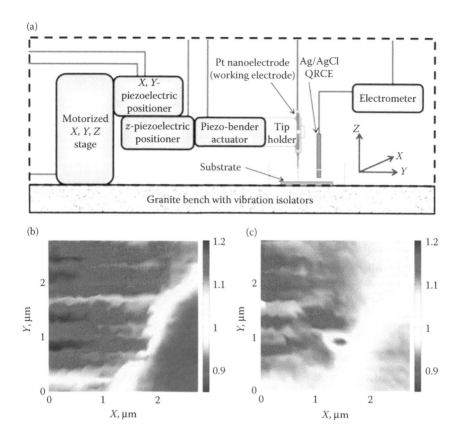

FIGURE 1.6 (a) Setup of intermittent-contact SECM with a piezoelectric bender actuator. (b) Forward and (c) reverse scan images of the edge of a gold band on glass obtained with a 160-nm-radius recessed Pt tip in 1 mM FcTMA+. (Reprinted from Lazenby, R. et al., *J. Solid State Electrochem.*, Vol. 17, 2013: pp. 2979–2987.)

substrate upon their contact. To prevent the damage problem in the intermittent-contact mode, a nanotip exerted a low force on the substrate by using a piezoelectric bender actuator with a low spring constant of 0.02 N/μm, which corresponded to a force of 1×10^{-4} N for the damping of tip oscillation by 5 nm. An even lower force might be exerted on the substrate owing to tip bending. Nevertheless, this careful study demonstrated that repetitive tip–substrate contacts resulted in the damage of a substrate. No damage was seen in the feedback image of a gold band array on glass as obtained by forward line scans of a 160-nm-radius Pt tip at a constant distance of 39 nm from the substrate (Figure 1.6b). Image contrast represents positive and negative feedback effects from the gold and glass surfaces, respectively, on the tip current based on the oxidation of (ferrocenyl-methyl)trimethyl ammonium (FcTMA+). In contrast, the image based on reverse line scans showed the feature of a negative feedback effect at $x = 1.5$ μm and $y = 1$ μm (Figure 1.6c), where a positive feedback effect was seen in the forward scan. With the exception of this feature, the forward and reverse scan images were very similar, while the resolution of the reverse scan image was lowered by employing a longer distance of ~300 nm from the forward scan z position to avoid a tip–substrate contact.

1.2.4 BELOW 100 NANOMETER RESOLUTION

Truly nanoscale lateral resolution of 1–100 nm has been reported for SECM imaging. Similarly high or even higher resolution is achievable by using other imaging methods, for example, electron microscopy, super-resolution fluorescence microscopy, and various scanning probe microscopy techniques such as AFM and scanning tunneling microscopy (STM). Uniquely, SECM offers this extremely high resolution for *in situ*, real-time, and quantitative imaging in solution. A combination of these features is advantageous for the study of biological systems as demonstrated by the topography imaging of biological macromolecules and cellular nanostructures. Moreover, quantitative nanoscale SECM imaging was employed to determine both the permeability and the size of single artificial nanopores as the model of biological nanopores as well as the electrocatalytic activity of single nanoparticles. The reactivity and topography imaging of a single nano-entity was also demonstrated with 9–100 nm resolution by using combined SECM techniques (Section 1.5).

1.2.4.1 Single Biological Macromolecules

Fan and Bard demonstrated the highest lateral resolution of ~1 nm for SECM in topography imaging of single biological macromolecules in the constant-current mode [36]. This study employed an STM-like setup with the exception that target macromolecules were adsorbed on the mica surface covered with an extremely thin water layer (Figure 1.7a). A sharply etched tungsten wire served as a nanotip without insulation because a redox-active species was electrolyzed only at the portion of the tip end immersed in the thin water layer. The resultant current of ~1 pA at the tip with a bias of 1 V against the large Au counter electrode corresponds to a contact radius of <3 nm. Since the average thickness of the water layer on mica was only subnanometers, the tip current was limited by a negative feedback effect from the mica and the target macromolecule to enable topography imaging in the constant-current mode. Figure 1.7b shows the high-resolution image of several DNA molecules on mica as obtained using a sharp tungsten tip biased at 3 V at 80% relative humidity. Noticeably, such high-resolution images of DNA on mica were originally reported by Guckenberger et al. [37], who ascribed the picoampere tip current to electron tunneling in the STM mode. It, however, was demonstrated later by Bard and coworkers that the tip current involved faradic current and that the resultant image was based on SECM rather than STM [38–39]. In addition to DNA, remarkably high resolution was demonstrated for SECM imaging of other biological macromolecules such as antibody, enzyme (e.g., glucose oxidase), and keyhole limpet hemocyanin molecules [36]. Figure 1.7c shows an image of mouse monoclonal immunoglobulin G (IgG) molecules on the mica surface. Importantly, the SECM images of native IgG molecules are similar to those found with x-ray crystallography and transmission electron microscopy (TEM). More quantitatively, the apparent dimensions of the IgG

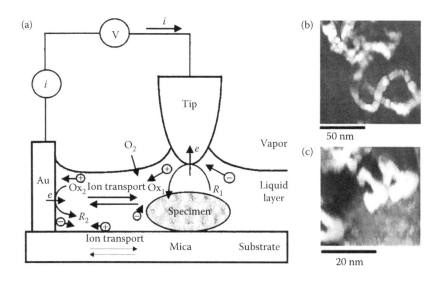

FIGURE 1.7 (a) Schematic diagram for the SECM setup with controlled humidity, where the electrochemical processes control the current. Constant-current image of (b) fragments of DNA specimen and (c) IgG molecules in humid air. The mica substrate was treated with a buffer solution. (Reprinted with permission from Fan, F.-R. F. and A. J. Bard, *Proc. Natl. Acad. Sci. USA*, Vol. 96, 1999: pp. 14222–14227.)

molecules (~12 nm × 9 nm) in the SECM image are significantly larger than those determined from x-ray crystallography (8.5 nm × 6 nm). The approximate height of the antibody molecule in the SECM image is 5 nm, which is somewhat higher than the TEM and x-ray data (~4 nm). These differences may be due to resolution limitation by tip size, while the size of a moist isolated molecule may be different from that in crystals. Noticeably, the requirement of TEM and x-ray data for the interpretation of the SECM image indicates its remarkably high resolution. By contrast, high-resolution SICM only detects single protein molecules [40] and channels [41] without resolving any structural detail.

1.2.4.2 Single Nanopores

A spatial resolution of ~30 nm was recently demonstrated by employing the constant-height mode to quantitatively image ion transport through the single nanopores of a porous nanocrystalline silicon (pnc-Si) membrane [42]. The ability of SECM to assess single pore permeability in liquid is important because pnc-Si membranes have found applications in liquid environments for macromolecular filtration, nanoparticle separation, cell culture, artificial kidney, electroosmotic pump, etc. [43–44]. In this SECM study, a nanopipette was filled with a DCE solution of organic supporting electrolyte and was immersed in an aqueous solution to form a nanometer-sized ITIES at the tip end (Figure 1.8a). Two metal electrodes were positioned inside and outside of the nanopipette to control the phase boundary potential across the ITIES. The interfacial transfer of an aqueous target ion into the organic phase was driven potentiostatically to generate an ionic current response. In SECM imaging, this tip current was lowered when the nanopipette was positioned within a tip diameter from the impermeable region of the membrane, which hindered the diffusion of the target ion to the nanoscale ITIES. In comparison to the negative feedback current, a higher tip current was obtained when the tip was positioned at the same height over a nanopore as an ion source.

Figure 1.8b,c shows the TEM image of a pnc-Si membrane and the constant-height SECM image of ion transport through single silicon nanopores, respectively [42]. Approximately 34-nm-diameter pipette tip was positioned at a distance of down to 1.3 nm from the membrane by monitoring the tip current response to

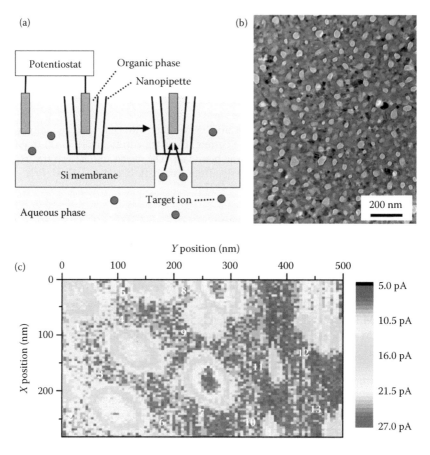

FIGURE 1.8 (a) Scheme of an SECM line scan with a nanopipette-supported ITIES tip over the impermeable and nanoporous regions of the membrane. (b) TEM image of a pnc-Si membrane. (c) SECM image of a pnc-Si membrane in 10 mM tetrabutylammonium and 0.3 M KCl obtained by using the ~34-nm-diameter nanopipette tip filled with a DCE solution of supporting electrolyte. (Reprinted with permission from Shen, M. et al., Quantitative imaging of ion transport through single nanopores by high-resolution scanning electrochemical microscopy, *J. Am. Chem. Soc.*, Vol. 134, 2012: pp. 9856–9859. Copyright 2012, American Chemical Society.)

tetrabutylammonium and then scanned laterally above the flat membrane at the same height. Higher current was observed over 13 nanopores in the 280 nm × 500 nm image as expected from a density of 90 nanopores/μm^2 in the TEM image of the membrane. The SECM image of nanopore 7 agreed very well with the theoretical image simulated by using the finite element method to yield major and minor axes of 53 and 41 nm and a depth of 30 nm for the nanopore as an elliptical cylinder. These lateral pore dimensions were smaller than those in the SECM image by the tip diameter, which determined the lateral resolution of the image. This spatial resolution of ~34 nm was high enough to resolve the dense silicon nanopores with a short pore–pore separations of only ~100 nm. Such high-density nanopores have not been resolved by using the other electrochemical imaging methods such as SECM [45–52], SICM [53–58], SECM–SICM [59–60], and SECM–AFM [61–63]. In these previous imaging studies, the shortest separations between two resolvable pores were limited to >250 nm and ~1.5 μm for SECM(–AFM) [62] and SICM [58], respectively. Further development of high-resolution SECM imaging will enable us to image the intrinsic structure and transport properties of single biological nanopores in water without chemical fixation or physical contact [64].

1.2.4.3 Single Biological Cells

Mirkin and coworkers are the first not only to demonstrate nanoscale SECM imaging of single biological cells but also to achieve a spatial resolution of ~100 nm [65]. In this study, $Ru(NH_3)_6^{3+}$ was employed as a membrane-impermeable redox probe to obtain the topographic image of the cell membrane in the negative feedback mode. Figure 1.9a shows the constant-current topography image (11.8 μm × 11.2 μm) of the cell surface as obtained with a 123-nm-radius Pt tip. This Pt tip was sealed in a thin-wall glass sheath with a small RG value of <3 to achieve a short distance between the tip and the cell membrane, thereby yielding the spatial resolution limited by the tip diameter. An even smaller tip (47 nm radius) was employed to obtain the constant-height image of a smaller and flatter region of the cell surface (Figure 1.9b), where no topographic feature or reactive site was clearly visible. This pioneering work was followed by Bergner et al. [66], who obtained reactivity information by constant-height imaging of biological cells although spatial resolution was limited to ~600 nm. Specifically, ~0.3-μm-radius Pt tip was employed to study passive membrane transport of FcMeOH as a membrane-permeable probe at the confluent monolayer of epithelial cells. Such a Pt nanotip was also used for triggering the necrosis of an individual cell within the cell monolayer by locally generating hydroxide ions [67].

Remarkably, the tip radius of a nanoelectrode can be ~1000 times smaller than the size of a cell so that the nanotip can penetrate through the cell membrane for direct characterization of the intracellular redox state [65]. The penetration of a nanotip into the cell was monitored as a change in the amperometric tip current as the tip moved perpendicularly to the cell surface, that is, approach curve. The tip current based on FcMeOH oxidation was constant while the tip was moving outside of the cell (Figure 1.9c). As the tip approached within a feedback distance from the cell, tip current decreased because the diffusion of FcMeOH to the tip was

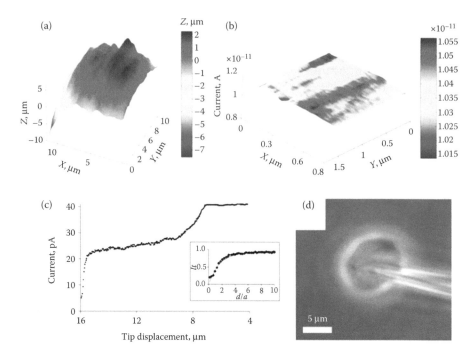

FIGURE 1.9 (a) Constant-current and (b) constant-height images of a human breast epithelial (MCF-10A) cell obtained with 123- and 47-nm-radius Pt nanotips, respectively. (c) Approach and penetration of an MCF-10A cell by a 144-nm-radius Pt tip. The inset shows the experimental (symbols) and theoretic (solid line) curves for the final part of the approach curve, where the tip approached the bottom of the petri dish. The extracellular PBS solution contained 1 mM $Ru(NH_3)_6Cl_3$ in (a) and 1 mM $FcCH_2OH$ in (b) and (c). (Reprinted with permission from Sun, P. et al., Nanoelectrochemistry of mammalian cells, *Proc. Natl. Acad. Sci. USA*, Vol. 105, 2008: pp. 443–448.) (d) Optical micrograph of a nanotip ($a = 75$ nm; 800 nm OD) inside a macrophage. (Reprinted with permission from Wang, Y. X. et al., *Proc. Natl. Acad. Sci. USA*, Vol. 109, 2012: pp. 11534–11539.)

hindered by the cell membrane. Tip current, however, decreased only to ~70% of its value in the bulk solution because the cell membrane was permeable to FcMeOH. In fact, tip current reached to a lower plateau value as the tip penetrated through the cell membrane to detect intracellular FcMeOH molecules. The lower current is likely due to slower diffusion of FcMeOH in the viscous cytoplasm. Finally, a pure negative feedback was seen as the tip approached toward the cell membrane on the petri dish. The approach curve in this regime agreed with the theoretical negative approach curve (the inset of Figure 1.9c), thereby confirming the quantitative electrochemical response of the Pt nanotip inside the cell. In contrast, a less negative feedback response to membrane-permeable menadione was obtained by approaching a Pt nanotip to the cell membrane on an Au substrate, which regenerates menadione. The approach curve was analyzed to determine a high membrane permeability of ~0.15 cm/s. More recently, platinized Pt nanotips were developed to detect reactive oxygen and nitrogen species inside a macrophage (Figure 1.9d) [68].

A platinized nanotip served also as an excellent potentiometric probe to enable the measurement of intracellular redox potentials [69].

Nanoscale constant-current imaging was demonstrated for various types of biological cells with unique topographic features [29]. This work also employed $Ru(NH_3)_6^{3+}$ as a membrane-impermeable hydrophilic probe to obtain the topographic image of a cell surface in the negative feedback mode. Carbon-based nanotips with a small RG value of ~3 were operated in the hopping mode (Figure 1.5a) to enable noncontact topography imaging of micrometer- and submicrometer-sized cellular features. For instance, the boundary area between the differentiated PC12 cell and the substrate surface showed a steep slope in the topography image as obtained by using an 18.9-nm-radius carbon nanotip (Figure 1.10a). The topography image of cardiac myocyte (Figure 1.10b) was also obtained using a 12.0-nm-radius tip to visualize wave structures. Moreover, the SECM topography image of hippocampus neurons with a 32.6-nm-radius tip showed the protrusions of synaptic boutons, the swollen end of an axon that contributes to a synapse (Figure 1.10c). This SECM topography image was consistent with the confocal fluorescence microscopic image of the same region of the hippocampal neurons as stained with FM 1-43 for labeling of synaptic vesicles. Noticeably, the size of the nanopipette tips filled with pyrolytic carbon is likely underestimated as discussed in Section 1.4.1.3.

1.2.4.4 Single Nanoparticles

Recently, a remarkably high spatial resolution of down to ~6 nm was demonstrated by imaging the electroactivity of individual Au nanoparticles [70]. This high spatial resolution was achieved by employing a 3.1-nm-radius Pt tip in the feedback mode with FcMeOH as a redox mediator (Figure 1.11a). In this experiment, 10-nm-diameter Au nanoparticles were sparsely dispersed on a thin insulating polyphenylene film coated on highly oriented pyrolytic graphite (HOPG). In the feedback mode, FcMeOH was oxidized at the tip and was regenerated only at an Au nanoparticle partially buried into the insulating film to show the

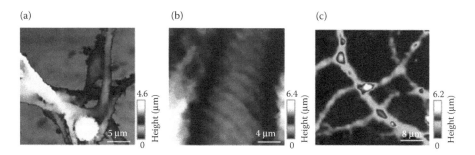

FIGURE 1.10 Constant-current topography images of (a) differentiated PC12 cells, (b) cardiac myocyte, and (c) hippocampus neurons obtained with the carbon-based nanotips held at −500 mV versus Ag/AgCl in PBS containing 10 mM $Ru(NH_3)_6Cl_3$. (Reprinted with permission from Takahashi, Y. et al., *Proc. Natl. Acad. Sci. USA*, Vol. 109, 2012: pp. 11540–11545.)

feedback enhancement of the tip current over a single nanoparticle (Figure 1.11b). Importantly, the size of the nanoparticle in the SECM image was consistent with the actual size of the nanoparticle. The lateral resolution in this image was significantly higher than that in AFM images as obtained with typical commercial probes, which caused an artifact due to the tip convolution effect. In fact, a lateral nanoparticle size of ~50 nm was obtained by AFM imaging of ~20-nm-diameter nanoparticles on the insulated HOPG surface. In addition, the steady-state feedback effect from the nanoparticle on the unbiased HOPG electrode indicated the electrical connection between the nanoparticles and the underlying electrode. This connection was good enough to study H$^+$ reduction at individual nanoparticles in the SG/TC mode (Figure 1.11c). Specifically, a 15-nm-radius Pt tip was scanned at

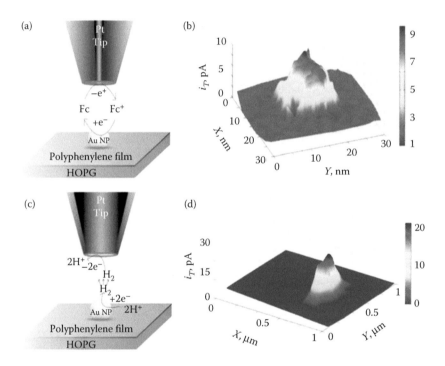

FIGURE 1.11 (a) Schematic representation of the feedback mode at a single nanoparticle and (b) a constant-height SECM image of a 10-nm-diameter Au nanoparticle as obtained with a 3.1-nm-radius Pt tip. The solution contained 1 mM FcMeOH and 0.1 M KCl. The tip potential was 400 mV versus Ag/AgCl and the HOPG substrate was unbiased. (c) Schematic representation of the SG/TC experiment at a single NP and (d) a constant height image of the hydrogen evolution reaction at a single 20-nm-diameter Au nanoparticle as obtained with a 15-nm-radius Pt tip. The tip and substrate potentials were 500 mV and −750 mV versus Ag/AgCl, respectively. The solution contained 10 mM HClO$_4$ and 0.1 M NaClO$_4$ (From Sun, T. et al., Scanning electrochemical microscopy of individual catalytic nanoparticles, *Angew. Chem., Int. Ed.*, 2014. Vol. 53. pp. 14120–14123. Copyright 2014, Wiley-VCH Verlag GmbH & Co. KGaA. Reproduced with permission.)

80 nm from a 20-nm-diameter Au nanoparticle to oxidatively detect H_2 generated at the nanoparticle, which was biased through the HOPG electrode. The spatial resolution of the SECM image is much lower than the tip size (Figure 1.11d) owing to diffusional broadening across the tip–nanoparticle gap, which is much wider than the tip size. Quantitative information about nanoscale H_2 evolution and oxidation was also obtained by voltammetrically studying the tip current as a function of nanoparticle and tip potentials, respectively. A voltammogram based on the tip current versus the nanoparticle potential gave a slope of 116 mV per decade for the Tafel plot of H_2 evolution at the single nanoparticle. This slope is consistent with literature date for polycrystalline gold. On the other hand, a Tafel slope of 126 mV per decade was obtained for H_2 oxidation at the Pt nanotip from a voltammogram based on the tip current versus the tip potential.

1.3 NANOGAP EXPERIMENTS

The nanometer-wide gap formed between the tip and the substrate is useful for SECM experiments beyond nanoscale imaging. Extremely narrow gaps with a width of 1–10 nm have been used to electrochemically detect the single molecules trapped within the nanogaps [71]. The SECM-based nanogaps with a width of 1–100 nm have been also employed for the kinetic study of fast charge-transfer reactions at the tip or the substrate as well as fast homogeneous reactions in the gap. In comparison to the micrometer-scale counterpart, the SECM-based nanogap approach offers high mass transport conditions not only to address faster kinetic regimes but also to widen the range of examinable experimental conditions. In addition, nanogap-based SECM is advantageous against lithographically fabricated nanogap electrodes [72], which are also useful for single-molecule detection [73] and fast kinetic study [74]. In principle, an SECM-based nanogap can be formed over any substrate including solids, liquids, membranes, and potentially gases. Moreover, an SECM-based nanogap can be significantly narrower than a nanogap with a width of down to 30 nm as formed by the nanolithographic approach [73]. The width of an SECM-based nanogap is precisely adjustable by monitoring feedback current at the tip, which is highly sensitive to the tip–substrate distance [75].

1.3.1 SINGLE MOLECULE DETECTION

The exciting application of an extremely narrow SECM-based nanogap is the electrochemical detection of single molecules [71]. Current measurement with presently available instrumentation requires several thousands of electron-transfer events at an electrode, which corresponds to zeptomoles of molecules [76–77]. In contrast, the positive feedback effect of SECM amplifies a current response to a single molecule to enable its electrochemical detection [78]. Specifically, a small volume of a dilute solution of an electroactive species was trapped in the cavity of a 15-nm-diameter Pt-Ir tip surrounded by insulating wax sheath over indium-tin oxide as a conductive substrate (Figure 1.12a). The tip–substrate distance was adjusted to ~10 nm by monitoring the positive feedback current response to the

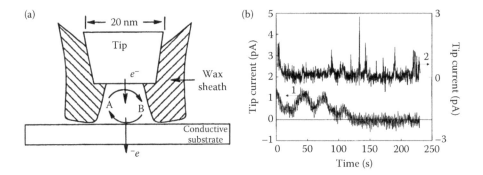

FIGURE 1.12 (a) Electrochemical detection of a single molecule, A, trapped between the tip and the substrate. (b) Tip current with time for (curve 1) a solution of 2 mM FcTMA$^+$ and 2.0 M NaNO$_3$ with ~10-nm-radius Pt-Ir tip at 0.55 V versus SCE and an ITO substrate at −0.3 V and (curve 2) a solution containing only 2.0 M NaNO$_3$ with ~7-nm-radius Pt-Ir tip within the tunneling range from the substrate. Data sampling rate 0.4 s per point. (Reprinted with permission from Bard, A. J. and F.-R. F. Fan, Electrochemical detection of single molecules, *Acc. Chem. Res.*, Vol. 29, 1996: pp. 572–578. Copyright 1996, American Chemical Society.)

electroactive species at the tip. Importantly, the time required for a single mediator molecule to travel across a gap, τ, varies with the square of the gap width, d, as given by

$$\tau = \frac{d^2}{(2D)} \tag{1.2}$$

where D is the diffusion coefficient of the mediator molecule in the gap. Thus, an extremely short traveling time of 50 ns is expected for a narrow gap with $d = 10$ nm from Equation 1.2 with $D = 1 \times 10^{-5}$ cm^2/s. Subsequently, a single molecule can cycle the nanogap 10^7 times per second to provide the 10-million-fold amplification of the current response when each cycle results in the transport of one electron between the tip and the substrate. The resulting current of 1.6 pA is measurably high to enable the electrochemical detection of the single molecule. In fact, such a current response was observed amperometrically for the single molecule of FcTMA$^+$ trapped within a 10-nm-wide tip–substrate gap (line 1, Figure 1.12b). The fluctuation of the tip current was observed as the single molecule moved into and out of the tip–substrate gap. Importantly, the stochastic faradic current response to the single molecule was much broader in the time series than tunneling current between the tip and the substrate (line 2 in Figure 1.12b). The tunneling current response was observable without an electroactive species in the solution when the tip–substrate gap was narrowed down to a tunneling distance. Furthermore, statistical methods were employed to confirm that the fluctuation amplitude of the tip current agreed well with the current expected for a single molecule in the thin-layer cell of the geometry shown in Figure 1.12a [79].

Interestingly, an SECM-based nanogap can be also used to detect the emission of electrogenerated chemiluminescence (ECL) from a single molecule [71].

Recently, a nanogap was formed between a recessed Pt nanotip and a Hg pool to enable the steady-state electrochemical detection of single or a few electroactive molecules without their escape from the gap (Figure 1.13a) [80]. With this approach, the cavity of the etched Pt nanotip was filled with the aqueous solution of the redox mediator and immersed into the dry Hg pool to form a sealed thin-layer cell. The depth and radius of the tip cavity were well characterized by approach curve measurement [81] to be consistent with a limiting current as obtained when a large number of redox molecules were trapped in the thin-layer cell formed by a relatively large tip (~100 nm radius). When a much smaller recessed tip with a radius of 15.2 nm was employed to repeatedly form a 6.8-nm-thick cell, the resultant liming current based on FcMeOH oxidation varied randomly (Figure 1.13b). The limiting currents corresponded to the values expected for the small number of molecules, that is, 0, 1.1, 2.1, 3.0, and 3.6. The number of randomly trapped FcMeOH molecules was distributed around an average number of 2.96 molecules as expected for the corresponding number of FcMeOH molecules (1 mM) in the zeptoliter volume of the nanoscale thin-layer cell. Importantly, the molecules were trapped within the confined volume to give steady-state voltammograms without the large fluctuation of tip current, which contrasts to the stochastic current response observed with the leaky thin-layer cell (Figure 1.12b) [78–79]. This steady-state approach facilitated single-molecule studies under various experimental conditions including different redox species and different concentrations of supporting electrolytes. These detailed studies revealed such unusual electrochemical phenomena as current rectification and pinning of Hg potential.

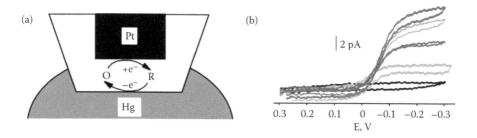

FIGURE 1.13 (a) Schematic representation of an etched Pt nanotip and the thin water layer trapped inside the glass sheath when the recessed nanotip was immersed into the pool of Hg. (b) Steady-state voltammograms obtained by random sampling of single molecules using a 15.2-nm-radius Pt tip in a 6.8-nm-thick cell. The voltammograms were obtained by repeatedly transferring the etched Pt nanotip from 1 mM $FcCH_2OH$ solution to a mercury pool. The limiting currents were used to calculate the number of FcMeOH molecules, N, that is, 0, 1.1, 2.1, 3.0, and 3.6 from the bottom to the top. (Reprinted with permission from Sun, P. and M. V. Mirkin, Electrochemistry of individual molecules in zeptoliter volumes, *J. Am. Chem. Soc.*, Vol. 130, 2008: pp. 8241–8250. Copyright 2008, American Chemical Society.)

1.3.2 REACTION KINETICS

Mediating fast charge-transfer reactions at interfaces is the central theme of electrochemistry with both fundamental and practical importance. The kinetic measurement of fast interfacial reactions, however, is often compromised by the diffusion of a reactant and a product between the solution and the interface. This limitation can be alleviated by employing SECM-based nanogap approach, which offers unprecedentedly high mass transport conditions to enable the study of fast heterogeneous and homogeneous reactions. In fact, the highest standard electron-transfer rate constants, k^0, reported so far were determined by the SECM-based nanogap approach. A highly reactive intermediate with a microsecond lifetime was also detected with the SECM-based nanogap configuration. These fast reactions were studied by using nanogaps with truly nanoscale widths of 1–100 nm.

1.3.2.1 Advantages and Limitations

The advantages and limitations of kinetic measurement with SECM-based nanogaps can be quantitatively evaluated by considering a simple relationship between mass transfer coefficient, m, and gap width, d, as approximately given by Reference [82]

$$m = \frac{D}{d} \qquad (1.3)$$

In principle, a narrower gap gives a higher mass transfer coefficient to enable the kinetic measurement of a faster reaction. An exceptionally high m value of 10 cm/s is expected from Equation 1.3 with $d = 10$ nm and $D = 1 \times 10^{-5}$ cm²/s. A heterogeneous rate constant of up to 100 cm/s (=10m) is measurable without diffusion limitation under such a high mass transport condition [83]. Remarkably, a mass transfer coefficient of 10 cm/s is not obtainable by using a rotating disk electrode, which must be rotated at 2.5×10^7 rpm to yield this high mass transfer coefficient. A high mass transport condition across a nanogap can be also characterized by the time required for a molecule to diffuse across the nanogap, τ, as given by Equation 1.2. The τ value corresponds to the shortest lifetime of the unstable product of a tip (or a substrate) reaction that can be detected at the substrate (or a tip), for example, $\tau = 50$ ns with $d = 10$ nm and $D = 1 \times 10^{-5}$ cm²/s in Equation 1.2. The nanosecond time scale is also achievable by cyclic voltammetry with an extremely fast scan rate of 500 kV/s, which requires a highly sophisticated and tailor-made potentiostat [84]. In contrast, the nanogap-based SECM detection of an unstable intermediate with a half-life of a few nanoseconds can be done at steady states by using a commercial bipotentiostat. Advantageously, a steady state across a nanogap is quickly reached in the positive feedback mode within a timescale of only $\tau/2$ at $d = 0.1a$, for example, 25 ns with a 10-nm-wide gap as predicted by the theory of SECM chronoamperometry [31]. Similarly, a quasi-steady state is quickly achieved within a nanogap when a substrate reaction is transient [17]. Significantly, the quasi-steady-state approach is required for the study of a reaction at a macroscopic substrate within a wide range of applied potential around a formal potential in both feedback and SG/TC modes (Section 1.3.2.3).

Practically, the width of an SECM-based nanogap is limited by contact between the tip and the substrate owing to their imperfect alignment. When a flat disk-shaped tip is tilted against a flat substrate with an angle, θ, the edge of the insulating sheath of the tip contacts the substrate surface at a distance, $d_{contact}$, as given by Reference [85]

$$\frac{d_{contact}}{a} = RG \tan \theta \qquad (1.4)$$

When the tip–substrate alignment is carefully adjusted [83], a tip with a typical RG value of 10 can approach to $d_{contact} < 0.1a$, which corresponds to $\theta < 0.6°$ in Equation 1.4. With such a good alignment, a 100- to 1000-nm-wide gap can be formed by using ~1-μm-radius tip with RG of ~10 [86–87]. Alternatively, Equation 1.4 predicts that a submicrometer-wide gap can be formed by using a much larger 5-μm-radius tip with $RG = 2$. Recently, this prediction was confirmed experimentally [20,88–91]. Narrower gaps with a 10–100 nm width have been also formed by using flat Pt [83] and Au [92] nanotips with $a = 5$–400 nm and $RG = 3$–10 over a flat Au surface. By contrast, the formation of metal–metal nanogaps with 1–10 nm width usually requires the cavity of a recessed metal nanotip with deformable insulating sheath [78] or with liquid mercury as a substrate [80] as discussed for single-molecule detection in Section 1.3.1. A nanogap with a width of <10 nm can be rarely formed by using mechanically polished Pt nanotips even with a radius of down to 3.7 nm [83]. The origin of this limitation is not well understood.

Eventually, SECM experiments with an extremely narrow tip–substrate gap can be limited by the observation of tunneling current, which becomes significant and may even exceed faradic current when a conductive nanotip is positioned within ~1 nm from a conductive substrate. This limitation has been discussed theoretically [93–94] and also observed experimentally [5] (see also Figure 1.12b [78]). Moreover, significant tunneling current seen from the rough surface of a tip [83] (or a substrate [17]) was positioned in the proximity of a substrate (or a tip). Noticeably, the control of a tip–substrate distance between feedback and tunneling regimes is crucial to combine SECM with STM [95–97] as a powerful tool for atomic-scale structure imaging and nanoscale reactivity imaging. In contrast to the metal–metal nanogaps, tunneling current is not a limitation when either tip or substrate is insulator or an ionic conductor. For instance, nanoscale SECM imaging with 1 nm resolution was enabled by scanning a sharp metal nanotip within ~1 nm from the biological macromolecules adsorbed on the insulating mica surface (Figure 1.7a) [36]. Another example is a nanopipette-supported ITIES tip, where the inner solution serves as an ionic conductor. ITIES-based nanopipette tips with small a and RG values of 10–20 nm and 1.5–2.0, respectively, were used to successfully form ~1-nm-wide gap in approach curve measurement (Section 1.4.1.2) [98] and imaging (Figure 1.8c) [42]. A nanopipette-based experiment also confirmed that the stability of a nanogap is seriously limited by thermal drift under ambient conditions (Section 1.4.4) [99]. Noticeably, the contamination of an SECM tip or substrate with organic impurities can be another limitation, which is manifested under high mass transport conditions across nanogaps to complicate the analysis of kinetic data. This contamination effect can be serious for nanotips, where organic impurities are efficiently transported from air or water to the small tip surface (see Section 1.4.3).

1.3.2.2 Tip Reaction

A fast electron-transfer reaction at a tip has been studied voltammetrically using a nanogap configuration in the positive feedback mode. Specifically, the steady-state voltammogram of the fast tip reaction is measured when the tip is positioned within a feedback distance form a conductive substrate to enhance mass transport across the nanogap. In the feedback mode, tip potential can be varied widely around E^0 to determine a k^0 value of up to 10 times of a mass-transfer coefficient, m [82], although determination of both k^0 and α from a steady-state voltammogram is less reliable when $k^0 > m$ [100]. Nanogap voltammetry of fast tip reactions was originally demonstrated by forming 100- to 1000-nm-wide gaps under ~1-μm-radius Pt tips to measure relatively large standard electron-transfer rate constants, k^0, of 0.12–0.46 cm/s for fullerene C_{60} [86] and ~3.7 cm/s for ferrocene [87] in acetonitrile. In the respective studies, a positive feedback effect was exerted from a mercury pool (Figure 1.10c) and a polished Pt macroelectrode. The flat and flexible surface of a mercury pool is advantageous for nanogap formation not only to trap single molecules (Section 1.3.1) but also to study their reaction kinetics as demonstrated for the fast reaction of single $Ru(NH_3)_6^{3+}$ molecules at a recessed Pt nanotip [80]. A remarkably high mass transfer coefficient of ~50 cm/s across a 1.2-nm-wide Pt-Hg nanogap resulted in a quasi-reversible tip voltammogram to yield an extremely high k^0 value of 8.6 cm/s and a normal α value of 0.58 (Figure 1.14a). Interestingly, a large number of $Ru(NH_3)_6^{3+}$ molecules were also studied voltammetrically at the flat Pt nanotips positioned at a

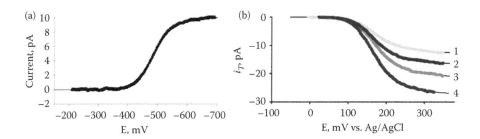

FIGURE 1.14 (a) Experimental voltammogram (symbols) of the single $Ru(NH_3)_6^{3+}$ molecule trapped in the 1.2-nm-depth cavity of a 8.2-nm-radius etched Pt electrode immersed in Hg. Solution contained 2 mM $Ru(NH_3)_6Cl_3$ and 0.2 M KNO_3. Theoretical curve (solid line) was calculated with $k^0 = 8.55$ cm/s and $\alpha = 0.58$. (Reprinted with permission from Sun, P. and M. V. Mirkin, Electrochemistry of individual molecules in zeptoliter volumes, *J. Am. Chem. Soc.*, Vol. 130, 2008: pp. 8241–8250. Copyright 2008, American Chemical Society.) (b) Experimental (symbols) and theoretical (solid lines) steady-state voltammograms of 1 mM FcMeOH and 0.2 M NaCl as obtained by positioning a 36-nm Pt tip over an Au substrate with $d = (1) \infty$, (2) 54, (3) 29, and (4) 18 nm. Potential sweep rate, 50 mV/s. (Reprinted with permission from Sun, P. and M. V. Mirkin, Kinetics of electron-transfer reactions at nanoelectrodes, *Anal. Chem.*, Vol. 78, 2006: pp. 6526–6534. Copyright 2006, American Chemical Society.)

distance of down to 15 nm from a flat Au substrate [83]. An even higher k^0 value of 17.0 cm/s was determined from quasi-reversible tip voltammograms to be the highest k^0 value reported so far for a heterogeneous charge-transfer reaction. A difference in the k^0 values determined by the respective approaches was ascribed to the difference of supporting electrolytes, that is, KNO_3 and NaCl.

Advantageously, the width of an SECM-based nanogap can be systematically changed to check whether reliable kinetic parameters are obtained independent of the gap width. This advantage was demonstrated by using Pt [83] and Au [92] nanotips with 15–300-nm-wide gaps formed over a flat Au substrate. With this approach, distance-independent kinetic parameters were obtained for various fast electron-transfer reactions including $Ru(NH_3)_6^{3+}$ (see above), FcMeOH, ferrocene, and 7,7,8,8-tetracyanoquinodimethane (TCNQ). For instance, Figure 1.14b shows a family of the tip voltammograms of FcMeOH as obtained by using a 36-nm-radius Pt tip at various distances of down to 18 nm from an Au substrate. Higher tip current was obtained at a shorter tip–substrate distance owing to a higher positive feedback effect from the conductive substrate. All voltammograms including other trials reproducibly gave k^0 values of 6.8 ± 0.7 cm/s and a normal α value of 0.42 ± 0.03. Similar kinetic parameters were also obtained with Au nanotips for most of the redox couples studied by Pt nanotips with the exception of the $Ru(NH_3)_6^{3+/2+}$ couple, which gave a significantly low k^0 value of 13.5 ± 2 cm/s at Au nanotips in comparison with Pt nanotips [92]. This difference revealed some degree of the nonadiabaticity of the $Ru(NH_3)_6^{3+/2+}$ couple, which has been considered to be adiabatic [101]. Noticeably, the power of nanogap-based tip voltammetry was also demonstrated for the kinetic study of fast electron-transfer reactions at the nanoscale ITIES supported at the tip of a nanopipette [102]. The utility of nanogap-based tip voltammetry, however, is limited because most materials and systems cannot be miniaturized as a tip to form a nanogap over a substrate. This limitation can be overcome by employing these materials and systems as macroscopic substrates rather than tips (Section 1.3.2.3).

Nanogap-based tip voltammetry was also enabled by using a 10-μm-diameter Pt tip with an extremely thin insulating sheath [103]. The glass sheath of the Pt tip was mechanically polished back away as seen in the SEM image (Figure 1.15a) [88]. The tip geometry and achievable tip–substrate distances were thoroughly characterized by studying the approach curves based on the positive feedback effect from a Pt substrate. Remarkably, the tip current was enhanced by a factor of ~30 in comparison to its limiting current in the bulk solution before the tip contacted the substrate (Figure 1.15b). The approach curve fitted very well with the theoretical curve simulated by considering the exact shape of the glass sheath as determined from the SEM image. This analysis showed that the closest tip–substrate distance of ~110 nm corresponded to only 2.2% of the tip radius. Such a sharp micrometer-sized Pt tip was successfully applied for nanogap-based tip voltammetry of tris(2,2′-bipyridine)ruthenium (II) ($Ru(bpy)_3^{2+}$) in acetonitrile [103]. This redox molecule is not only practically important for ECL [104] but also fundamentally interesting because it can serve as either acceptor or donor of an electron. The oxidation of the ruthenium center was slow enough to obtain a family

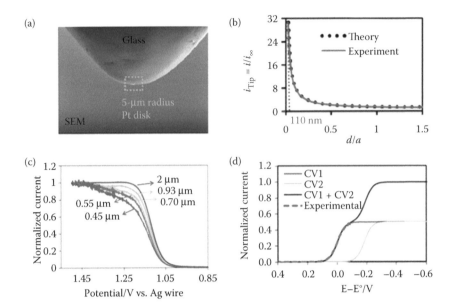

FIGURE 1.15 (a) SEM image of a 5-µm-radius Pt tip with an extremely thin-glass sheath and (b) a positive feedback approach curve obtained by carefully aligning the Pt tip over the highest point of a Pt substrate. (Reprinted with permission from Shen, M., and A. J. Bard, Localized electron transfer and the effect of tunneling on the rates of $Ru(bpy)_3^{2+}$ oxidation and reduction as measured by scanning electrochemical microscopy, *J. Am. Chem. Soc.*, Vol. 133, 2011: pp. 15737–15742. Copyright 2011, American Chemical Society.) Normalized $(i_T/i_{T,\infty})$ tip steady-state voltammograms for the (c) oxidation and (d) reduction of 0.38 mM $Ru(bpy)_3^{2+}$ in the acetonitrile solution of 0.1 M tetrabutylammonium hexafluorophosphate as obtained with a 5-µm-radius Pt tip with a thin-glass sheath. In panel C, distances, d, of 0.45–2 µm between the tip and the substrate are indicated. In panel D, the tip–substrate distance was 0.6 µm. Potential sweep rate, 0.001 and 0.005 V/s in panels C and D, respectively. (Reprinted with permission from Shen, M. et al., Achieving nanometer scale tip-to-substrate gaps with micrometer-size ultramicroelectrodes in scanning electrochemical microscopy, *Anal. Chem.*, Vol. 83, 2011: pp. 9082–9085. Copyright 2011, American Chemical Society.)

of quasi-reversible tip voltammograms with various nanogap widths (Figure 1.15c), thereby yielding consistent k^0 values of 0.7 ± 0.1 cm/s. In contrast, the reduction of the bipyridine ligand was too fast to observe a kinetic effect on a nanogap-based tip voltammogram. The reversible voltammogram was fitted well with the theoretical voltammogram based on the reversible reduction of the first and second ligands (CV 1 and CV 2, respectively, in Figure 1.15d) to estimate a minimum k^0 value of 3 cm/s for the first reduction. A higher k^0 value for bipyridine reduction indicates that this ligand serves as a barrier for the oxidation of the Ru center. Noticeably, nanogap-based tip voltammetry (and chronoamperometry) can employ arbitrary substrate potentials under steady or quasi-steady states to determine the diffusion coefficients [105] and short lifetimes [106] of both oxidized and reduced forms of a redox couple.

1.3.2.3 Substrate Reaction

An advantage of SECM resides in its versatility to study various types of electro-active materials and electrochemical systems as macroscopic substrates with their minimum modification. A reaction at a macroscopic substrate, however, must be relatively slow when its kinetics is studied in the steady-state positive feedback mode. In this operation mode, the substrate potential, E_S, must be far from the standard potential, E^0, so that the substrate reaction that is irreversible and opposite to the tip reaction occurs locally only under the tip (Figure 1.16a). Subsequently, the range of applicable substrate potential is significantly limited. Moreover, a fast substrate reaction is accelerated enough at E_S far from E^0 to be controlled by diffusion without reaching a kinetic regime. Thus, the largest k^0 value measured for a substrate reaction in the steady-state positive feedback mode has been limited to 0.42 cm/s for hydrogen oxidation at a macroscopic Pt substrate [107]. This moderately high k^0 value was obtained with ~300-nm-wide gap, which was formed by carefully aligning a 25-μm-diameter Pt tip over the substrate. This k^0 value was much

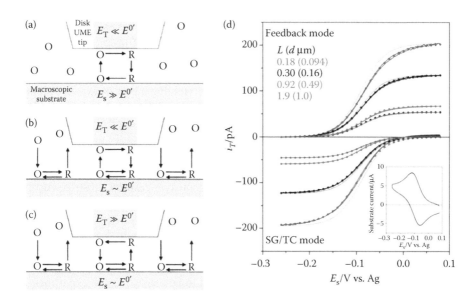

FIGURE 1.16 Scheme of nanogap-based SECM measurements of a fast electron-transfer reaction at a macroscopic substrate in the (a) steady-state feedback mode and in quasi-steady-state (b) feedback and (c) SG/TC modes. (d) Quasi-steady-state $i_T - E_S$ voltammograms of TCNQ in acetonitrile (solid curves). The tip was held at −0.235 or 0 V versus an Ag quasi-reference electrode for feedback or SG/TC modes, respectively. Substrate potential was cycled at 50 mV/s. Closed circles and dotted lines are theoretical curves for quasi-reversible ($k^0 = 7$ cm/s and $\alpha = 0.5$) and reversible substrate reactions, respectively, with $E^0 = -88$ mV. The inset shows a reversible voltammogram with a peak separation of 61 mV simultaneously measured at the substrate. (Reprinted with permission from Nioradze, N. et al., Quasi-steady-state voltammetry of rapid electron transfer reactions at the macroscopic substrate of the scanning electrochemical microscope, *Anal. Chem.*, Vol. 83, 2011: pp. 828–835. Copyright 2011, American Chemical Society.)

lower than a mass-transfer coefficient of ~2.4 cm/s as expected from Equation 1.3 with $d = 300$ nm and $D = 7.1 \times 10^{-5}$ cm^2/s for H$^+$. The k^0 value, however, was barely measurable owing to the limited range of substrate potential.

The reactivity of a macroscopic substrate can be studied in a wide range of substrate potential across the standard potential under quasi-steady-state conditions by employing the SECM-based nanogap approach [17]. Advantageously, the substrate reaction is slowed at E_S around E^0 to be reversible or even quasireversible, which is suitable for kinetic measurement. In this range of substrate potential, a substrate reaction is driven at the whole surface of the macroscopic substrate to develop a time-dependent concentration gradient of a mediator (Figure 1.16b), which contrasts to the steady-state feedback measurement (Figure 1.16a). Importantly, the concentration gradient around the gap can be negligibly small when the tip is positioned within ~1 μm from the macroscopic substrate. The resultant amperometric tip current in the feedback mode is obtained under quasi-steady-state conditions. Moreover, the substrate reaction in this potential range can be studied in the SG/TC mode under quasi-steady-state conditions simply by switching the tip potential for the amperometric detection of the substrate-generated species (Figure 1.16c). This quasi-steady-state approach was employed to study the fast one-electron reduction of TCNQ at a 2.5-mm-radius Pt electrode by using a 0.53-μm-radius Pt tip with $RG = 1.9$. Substrate potential was cycled across E^0 while tip potential, E_T, was set at $E_T < E^0$ or $E_T > E^0$ to reduce TCNQ in the feedback mode or oxidize its anion radical in the SG/TC mode, respectively, thereby yielding a pair of $i_T - E_S$ voltammograms at each tip–substrate distance from 1 μm to 94 nm (Figure 1.16d). The sigmoidal shape of the voltammograms confirms that the tip current was measured under quasi-steady-state conditions. The quasi-steady-state $i_T - E_S$ voltammograms of fast TCNQ reduction at a macroscopic Pt substrate changed from reversible to quasi-reversible as the 0.53-μm-radius Pt tip approached within 160 nm from the substrate (Figure 1.16d). In contrast, the simultaneously measured substrate current gives a reversible cyclic voltammogram with transient peak-shaped responses (the inset of Figure 1.16d).

A pair of quasi-steady-state voltammograms is obtained in the feedback and SG/TC modes to enable more reliable determination of kinetic parameters even when the corresponding reaction is nearly reversible. A unique combination of a high k^0 value of 7 cm/s and a normal α value of 0.5 had to be used to fit the experimental voltammograms of TCNQ at a macroscopic Pt substrate with the theoretical ones (Figure 1.16d). This k^0 value is higher than k^0 values of 1.1 cm/s [83] and 2.9 cm/s [106] as determined by SECM-based nanogap voltammetry at Pt nanotips and microtips, respectively. This analysis also gave a unique E^0 value of −88 mV versus an Ag quasi-reference electrode. In addition, $D = 1.6 \times 10^{-5}$ for the anion radical of TCNQ and tip–substrate distances were determined from a pair of limiting currents in the feedback and SG/TC modes to confirm high mass transfer coefficients of $m > 1$ cm/s with $d < 160$ nm in Equation 1.3. Quasi-steady-state voltammograms both in the feedback and SG/TC modes can be readily analyzed by using approximate analytical equations. These equations can also implement non-Butler-Volmer kinetics such as the Marcus–Hush–Chidsey formula of heterogeneous ET reactions [108], which can be important under extremely high mass transport conditions

[93,109]. Noticeably, the advantages of a pair of the steady-state voltammograms based on forward and reverse processes of a nearly reversible reaction were originally demonstrated by voltammetry of ion transfer at nanopipette-supported ITIES [100,110].

Quasi-steady state conditions were unknowingly used for the study of an IT reaction at the macroscopic DCE/water interface by using nanopipette-supported ITIES tips [111–112]. In these studies, a nanopipette was filled with an aqueous K^+ solution and immersed in the DCE solution of dibenzo-18-crown-6 (DB18C6) to drive the interfacial complexation reaction at the nanoscale ITIES as given by

$$K^+(\text{water}) + \text{DB18C6(DCE)} \rightarrow [K^+\text{DB18C6}](\text{DCE}) \tag{1.5}$$

The nanopipette was positioned within a feedback distance from the macroscopic ITIES formed between the DCE phase and the underlying aqueous K^+ solution supported on an Ag/AgCl electrode for the application of external bias and for the improvement of mechanical stability. A positive feedback effect was observed when the tip-generated complex was dissociated instantaneously at the macroscopic ITIES to regenerate free DB18C6 at the diffusion-limited rate. The dissociation process was slowed down to exert less positive feedback effect when the potential of the macroscopic ITIES, E_S, was set around the formal potential, $E_S^{0'}$, of the K^+-transfer reaction (Equation 1.5). Eventually, the E_S value was shifted further to dominantly form the complexes to deplete free DB18C6 near the whole macroscopic ITIES. This time-dependent mass transport condition, however, gave an apparently steady-state current response at the tip because the concentration of free DB18C6 was nearly uniform within the <500-nm-wide gap formed between the nanopipette and the macroscopic ITIES. The approach curves thus obtained under quasi-steady-state conditions were fitted well with theoretical steady-state curves for irreversible reactions. This analysis gave high ion-transfer rate constants, k_f, of 0.3–1.9 cm for the forward reaction of Equation 1.5 at various potentials in a range of $-0.05\,\text{V} < E_S - E_S^{0'} < 0.025\,\text{V}$ while a diffusion limit was reached at $E_S - E_S^{0'} < -0.05\,\text{V}$. A plot of k_f versus E_S was linear, thereby yielding k^0 values of 0.7 ± 0.3 cm/s and α values of 0.56 ± 0.08. These kinetic parameters agree with those determined by nanopipette voltammetry [113]. This data analysis, however, is erroneous because an IT reaction cannot be irreversible around $E_S^{0'}$.

1.3.2.4 Homogeneous Reaction

A fast homogeneous reaction can be also studied by employing an SECM-based nanogap approach. For instance, the microsecond lifetime of the cation radical formed by the oxidation of guanosine was measured by SECM [114]. This oxidation process is important and has been linked to the radiative damage of DNA. Specifically, guanosine was voltammetrically oxidized at the tip so that the tip-generated cation radical was amperometrically detected at the substrate. In contrast to most SECM experiments with a macroscopic substrate, a micrometer-sized substrate electrode was used to reduce background current. A nanometer-wide gap was successfully formed by using 10-μm-diameter carbon fiber ultramicroelectrodes as tip and substrate electrodes, which were aligned face-to-face by running SECM in the feedback mode with ferrocene as a redox mediator [115]. The tip-generated radical species was detectable at the substrate when the tip approached within 1 μm from the substrate. Substrate current increased at a shorter tip–substrate distance, where more cation radicals can travel from the tip to the substrate within their lifetime. The radical species was detected at the substrate with $d \leq \sim0.20$ μm, which gives a lifetime of ≤40 μs from Equation 1.2 with $D = 4.5 \times 10^{-6}$ cm²/s for guanosine. Advantageously, this short lifetime was measured under steady state conditions without the need of a short-time measurement in contrast to fast scan cyclic voltammetry [84].

More recently, nanogap-based SECM approach was employed to detect short-lived Sn(III) intermediates during the reduction of Sn(IV) in bromide media [116]. This electrochemical system is crucial to the development of a tin–bromine redox flow battery based on Sn(IV)/Sn(II) as the anolyte and Br^-/Br_2 as the catholyte. Advantageously, the steady-state SECM measurement of the Sn(IV)/Sn(II) system was free from a transient response to the surface processes of adsorbed interme-diates, which was dominant at high scan rates in fast scan cyclic voltammetry. Specifically, Sn(III) intermediates were generated at the tip and detected at the substrate, that is, the tip generation/substrate collection mode (Figure 1.17a). The tip was scanned from +0.2 to −0.3 V to voltammetrically reduce Sn(IV) while a sub-strate potential of −0.1 V was positive enough to oxidize Sn(III) species without the oxidation of Sn(II). The anodic current collected at the substrate became higher as the tip was positioned closer to the substrate, that is, down to 600 nm (Figure 1.17b). A shorter tip–substrate distance allowed for the detection of more Sn(III) intermedi-ates, which was quickly removed by the disproportionation reaction, for example, $2Sn(III)Br_5^{2-} \rightleftharpoons Sn(IV)Br_5^- + Sn(II)Br_5^{3-}$. The dependence of correction efficiency

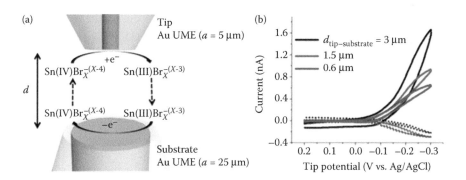

FIGURE 1.17 (a) Schematic diagram of the nanogap formed between two gold ultramicro-electrodes for the reduction of Sn(IV) at the tip and the oxidation of Sn(III) at the substrate in bromine media. (b) Tip voltammograms (solid line) and corresponding background-subtracted substrate current (dotted line) at various distances in 5 mM stannic bromide, 2 M HBr, and 4 M NaBr. (Reprinted with permission from Chang, J. and A. J. Bard, Detection of the Sn(III) intermediate and the mechanism of the Sn(IV)/Sn(II) electroreduction reaction in bromide media by cyclic voltammetry and scanning electrochemical microscopy, *J. Am. Chem. Soc.*, Vol. 136, 2014: pp. 311–320. Copyright 2014, American Chemical Society.)

on the tip–substrate distance was simulated numerically by considering the ECEC-DISP scheme. A good fit of the simulated result with the experimental one gave kinetic parameters for all steps including a disproportionation rate constant of 1.0×10^5/M/s. In another study, an even higher comproportionation rate constant of 1.0×10^6/M/s was determined for the consecutive two-electron reduction of TCNQ by using nanogap-based tip chronoamperometry [106].

1.4 ENABLING TECHNOLOGIES

Nanoscale SECM is technically challenging and has required the new development or application of several nanotechnologies. Challenges in nanoscale SECM originate from the fabrication, characterization, and use of a nanotip. Most of the challenges are common in electrochemical measurement with nanoelectrodes [117]. A nanotip can be seen as the nanostructured composite of the electrochemically active material surrounded by an inert material. The shape and size of the elaborate nanocomposite must be well controlled and characterized for its use in quantitative measurement. At the same time, the fragile nanotip must be interfaced to a macroscopic world for handling, characterization, and application without the damage of the nanotip or the compromise of its electroactivity. In addition, there are the challenges that are unique in nanoscale SECM to render it even more demanding. It is desirable for many SECM applications that the nanotip is comprised of the well-defined electro-active disk that is flush with the surrounding insulating sheath. The dimension and flatness of the electroactive disk and the surrounding insulator need to be controlled at nanometer or subnanometer scale. The insulating layer must be thin enough to enable the close approach of a nanotip to a substrate, which maximizes both positive and negative feedback effects. Moreover, the thin insulating sheath must be free from a pinhole, which compromises a feedback effect. A non-flat tip geometry also results in a small feedback effect [118]. Finally, accurate tip positioning with nanometer and subnanometer precision is crucial to nanoscale SECM.

1.4.1 NANOTIP FABRICATION

The reproducible fabrication of good nanotips has been facilitated by using a CO_2-laser puller and a focused ion-beam (FIB) instrument. The puller has been commercially available from Sutter to prepare a pair of nanopipettes with an outer tip diameter of down to 10 nm from a glass capillary with various compositions including quartz [119–121]. The nanopipette can be filled with a conductive material (either electronic or ionic) to serve as a glass-sealed or glass-supported nanotip. The robust glass-based nanotip can be mechanically polished to be smoothened and flattened. Alternatively, FIB milling enables the precise control of the shape and size of a nanotip. FIB technology is especially useful for the fabrication of the multifunctional nanotips elaborately designed for combined SECM techniques (Section 1.5.1).

1.4.1.1 Glass-Sealed Metal Nanotips

A metal nanotip is usually insulated by glass to take advantage of its mechanical robustness and chemical inertness in comparison to polymer materials, which were also examined in earlier studies [4,15]. Glass-sealed Pt nanotips have been prepared by heat sealing a sharply etched Pt wire into a glass capillary [4] or by heat pulling a glass-sealed Pt microwire by using a CO_2-laser puller [32,35,83,118,122–125] (Figure 1.18a). The latter approach is more popular because of simple and automatic procedures with the microprocessor-controlled puller and has been successful also for the fabrication of glass-sealed Au [92] and Ag [126] nanotips. Figure 1.18b–d shows steps for the preparation of a glass-sealed metal nanotip, that is, thinning a glass capillary, sealing a metal microwire in the capillary, and pulling the metal/glass composite [123]. All steps are carried out using the puller with the program based on five parameters, which are HEAT, FILAMENT, VELOCITY, DELAY, and PULL [119]. Typically, the heat-pulled metal tips have diameters of 10–1000 nm and RG values of 3–10. The RG values are small enough to facilitate the positioning of a nanotip within a feedback distance from a substrate. Remarkably, 1- to 3-nm-radius Pt electrodes were also obtained by heat pulling the Pt wire that was electrochemically etched and heat-sealed in a thick-wall glass capillary [127]. The molecularly small nanoelectrodes, however, are not suitable as an SECM tip because of their large outer diameters ($RG \gg 10$).

A glass-sealed Pt nanotip can be milled by FIB technology to obtain smooth and flat tip surfaces with diameters of down to 100–1000 nm [17,124,128] or recessed tips with diameters of 100–300 nm [32]. Advantageously, a single FIB platform offers both flexible nanomachining and high-resolution imaging [129]. In most commercially available systems, the beam of Ga^+ is scanned over a sample for its precise machining by the sputtering action of the energetic ion beam (1–30 keV). Secondary electrons are generated by the interaction of the ion beam with the sample surface and can be detected to obtain high-resolution images. In addition, most modern FIB

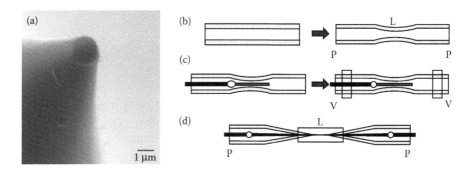

FIGURE 1.18   (a) SEM image of a 70-nm-radius polished Pt tip. (Reprinted with permission from Sun, P. and M. V. Mirkin, Kinetics of electron-transfer reactions at nanoelectrodes, *Anal. Chem.*, Vol. 78, 2006: pp. 6526–6534. Copyright 2006, American Chemical Society.) Protocol for the fabrication of a glass-sealed metal nanotip: (b) pulling (P) a glass capillary for its thinning by CO_2 laser (L); (c) insertion and sealing of a Pt wire in the capillary under vacuum (V); (d) pulling the glass-sealed metal wire. (Reprinted with permission from Mezour, M. A. et al. Fabrication and characterization of laser pulled platinum microelectrodes with controlled geometry, *Anal. Chem.*, Vol. 83, 2011: pp. 2378–2382. Copyright 2011, American Chemical Society.)

instruments supplement the FIB column with an additional SEM column. With this versatile "dual-beam" platform (FIB–SEM), the milled surface that is parallel to the Ga+ beam can be imaged by SEM. *In situ* SEM images in Figure 1.19 show the flat and smooth surfaces of Pt tips immediately after FIB milling [124]. In particular, FIB technology was useful for milling these Pt tips without damaging their thin glass sheath, which was obtained by annealing pulled tips [17,124,128]. Conveniently, the shape and size of a FIB-milled tip can be checked *in situ* by SEM to decide whether an additional milling step is needed to obtain a tip with a target size. Moreover, the final shape and size of a milled tip are determined by FIB and SEM imaging for quantitative analysis of electrochemical data. A disagreement of FIB and SEM images with electrochemical data revealed the unexpected damage of glass-sealed Pt nanotips [124] (Section 1.4.3).

A novel approach was recently proposed to develop the SECM nanotips based single Pt nanoparticles with diameters of down to 5 nm [130]. These nanotips were named "tunneling ultramicroelectrodes" because the tip current based on the electrolysis of a redox species at a nanoparticle is mediated by electron tunneling through a TiO_2 layer between the nanoparticle and an underlying submicrometer-sized Pt tip (Figure 1.20a). Specifically, ~200-nm-diameter Pt disk was exposed from a thin glass sheath by FIB milling as described above [128]. Then, a few nanometer thick TiO_2 film was electrodeposited on the exposed Pt surface to prevent the electrolysis of a redox mediator in solution. Importantly, the TiO_2 film was thin enough to mediate electron tunneling between the Pt electrode and the Pt nanoparticle adsorbed on the film because the high density of states (DOS) of the metal nanoparticle overlaps well with the DOS of the underlying Pt electrode. Subsequently, the electrolysis of a redox mediator at the nanoparticle was electrochemically detected at the Pt electrode. The adsorption of a single nanoparticle was not only confirmed by *ex situ* SEM but also monitored *in situ* by an amperometric current spike generated upon nanoparticle collision in a mediator solution (Figure 1.20b). Moreover, voltammograms and SECM approach curves at the nanoparticle-based tips agreed well with theory to yield an estimated nanoparticle diameter of down to 5 nm (Figure 1.20c). Significantly, the spherical tips gave substantial feedback effects both at conductive and at insulating

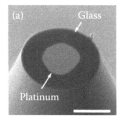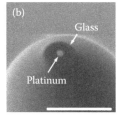

FIGURE 1.19 SEM images of the Pt tips sealed in borosilicate glass and milled by FIB. Tip diameters are (a) 940 and (b) 110 nm. Scale bars are 1 μm. (Reprinted with permission from Nioradze, N. et al., Origins of nanoscale damage to glass-sealed platinum electrodes with submicrometer and nanometer size, *Anal. Chem.*, Vol. 85, 2013: pp. 6198–6202. Copyright 2013, American Chemical Society.)

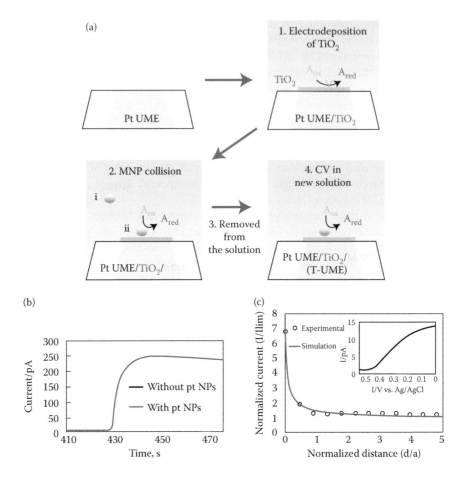

FIGURE 1.20 (a) Scheme of a TUME based on a single metal nanoparticle. (b) Chronoamperometric curve for the attachment of a Pt nanoparticle at the TiO$_2$-deposited Pt electrode in 10 mM K$_3$Fe(CN)$_6$ with (red) and without (black) 120 pM Pt nanoparticles. The amperometric tip current was acquired every 50 ms at −0.6 V versus Ag/AgCl. (c) Positive feedback approach curve at glassy carbon wafer in 10 mM K$_3$Fe(CN)$_6$ and 70 mM KCl (open circles) fitted with theory based on spherical Pt electrode geometry with 5 nm diameter (red line). The tip and substrate were held at −0.2 and 0.6 V versus Ag/AgCl, respectively. The inset shows a tip voltammogram in the bulk solution. (Reprinted with permission from Kim, J. et al., Tunneling ultramicroelectrode: Nanoelectrodes and nanoparticle collisions, *J. Am. Chem. Soc.*, Vol. 136, 2014: pp. 8173–8176. Copyright 2014, American Chemical Society.)

substrates, thereby indicating their high potential for high-resolution imaging and fast kinetic measurement.

1.4.1.2 Nanopipette-Supported ITIES Tips

A nanometer-scale ITIES can be supported by a nanopipette to serve as a versatile SECM tip to probe both electron-transfer and ion-transfer reactions [120].

A nanopipette can be filled with the solution of a redox active species to drive the electron-transfer reaction of a species in the outer solution [102]. More often, a nano-pipette-supported ITIES tip is used as an ion-selective tip as shown in Figures 1.4c [28] and 1.8a [42]. A nanopipette can be readily prepared by the CO_2-based puller and is filled with the organic or aqueous solution of supporting electrolytes. The thin glass wall of a nanopipette with a small *RG* value of 1.5–2.0 is highly advantageous for the close approach of the tip to a substrate. A nanopipette serves as either organic- or aqueous-filled tip. An organic-filled nanopipette was used to image a substrate in aqueous solution in the IT-based feedback mode (Figures 1.4 [28] and 1.8 [42]). The ion selectivity of an ITIES-based nanotip can be controlled by doping the inner organic solution with an ionophore [34,131]. An aqueous-filled nanopipette was immersed in a non-aqueous solution for IT-based nanoscale imaging in an ionic liquid [25] and also for the nanogap-based kinetic measurement of electron transfer at the tip [102] and ion transfer at the macroscopic ITIES [111–112]. Recently, metal nanoparticles (Ag, Au, and Pt) were electrodeposited at the ITIES supported by a water-filled nanopipette to fabricate metal nanotips for SECM [132].

Importantly, the hydrophilic wall of a glass nanopipette must be silanized appropriately to form a nanoscale ITIES at the very tip. The inner wall must be silanized and hydrophobic enough to prevent water from getting drawn into an organic-filled nanopipette. Originally, the inner wall of a nanopipette was silanized by dipping the pipette tip into chlorotrimethylsilane for 5–7 s as pretreatment for an organic-filled nanopipette [25,98]. In this case, the outer wall of the pipette was also silanized, but the inner organic solution did not form a layer on the hydrophobic outer wall [133]. The degree of pipette silanization strongly depends on silanization reagent, temperature, and humidity [134–135] and can be better controlled with vapor silanization [99]. With this approach, nanopipettes were first dried in the mini-vacuum desiccator that was evacuated by a pump [28]. Then, the dried pipettes were silanized by the vapor of highly pure and less reactive *N*-dimethyltrimethyl silylamine, which was delivered from a flask into the desiccator. Noticeably, the silanization of the outer pipette wall in the vapor of chlorotrimethylsilane also gave consistent results for the preparation of a water-filled nanopipette [110]. The outer wall of a water-filled nanopipette must be silanized to prevent water from spreading out to the outer wall. In contrast, dipping a nanopipette in liquid chlorotrimethylsilane often resulted in oversilanization to block the tip opening [136]. In both vapor and dipping methods, the silanization of an inner pipette wall was prevented by passing the flow of argon through the pipette from the back.

The roughness of a nanopipette tip can be reduced by mechanical polishing or by FIB milling to enable closer approach of the tip to a substrate. A borosilicate nanopipette was successfully polished by using a micropipette beveller without breaking the tip or plugging the orifice with polishing material (Figure 1.21a) [98]. SEM images clearly showed that a polished nanotip was smoother than an unpolished nanopipette tip (Figure 1.21b,c, respectively). In fact, the nanoscale ITIES tip supported by a polished nanopipette approached to an extremely short distance of ~0.8 nm from a substrate (1.4.2.2). Alternatively, the tip of a heat-pulled pipette can be smoothened by FIB milling [64,137]. FIB milling of a glass nanopipette with a diameter of down to 100 nm is possible [138]. The size of the smallest nanopipette that can be milled

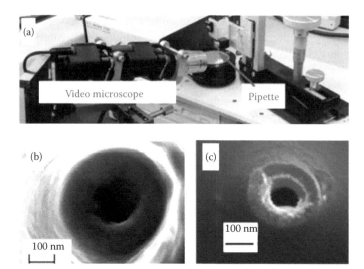

FIGURE 1.21 (a) Setup for nanopipette polishing. SEM images of (b) polished and (c) unpolished borosilicate nanopipettes. (Reprinted with permission from Elsamadisi, P. et al., Polished nanopipets: New probes for high-resolution scanning electrochemical microscopy, *Anal. Chem.*, Vol. 83, 2011: pp. 671–673. Copyright 2011, American Chemical Society.)

by FIB technology will be limited by the resolution of a FIB instrument, which can be compromised by the charging of the insulating glass surface. An additional disadvantage of FIB milling against mechanical polishing is a relatively high user fee.

1.4.1.3 Carbon-Based Nanotips

A carbon nanotip can be fabricated by the deposition of pyrolytic carbon in a glass nanopipette as originally proposed by Ewing and coworkers for the fabrication of carbon-ring ultramicroelectrodes [139]. In recent studies, pyrolytic carbon was deposited from butane gas under an argon atmosphere (Figure 1.22a) to fill either double barreled [140–141] or single barreled [29] nanopipette. In the former case, pyrolytic carbon can be deposited in both barrels (Figure 1.22b) [140] or in one barrel by plugging the other barrel to prevent butane flow (Figure 1.22c) [141]. The asymmetric double-barrel nanopipette is useful for combined SECM–SICM imaging (Section 1.5.3). A single carbon nanotip (Figure 1.22d) was employed for the nanoscale SECM imaging of single biological cells (Figures 1.5 and 1.10) [29]. Noticeably, the shape of the carbon nanotips is not well defined as indicated by *in situ* SECM characterization. For instance, a remarkably small carbon nanotip with a radius of down to 6 nm showed very small feedback effects from either insulating or conductive substrate [29], which is expected for a non-flat tip as discussed in Section 1.4.2. The small tip radius that was determined by assuming an inlaid disk-shaped tip was likely underestimated. Moreover, experimental approach curves with double-barreled carbon nanotips significantly deviate from theoretical curves, which suggests that the true tip geometry is recessed and larger than determined from the theoretical model (major axis size in a range of 100–400 nm) [140].

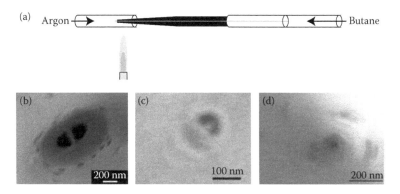

FIGURE 1.22 (a) Schematic of carbon deposition for fabrication of a dual carbon nanotip. Butane was passed through a pulled quartz theta pipette and pyrolyzed using a hand-held butane torch under an argon atmosphere. (b and c) Double- and (d) single-barrel nanopipettes thus prepared are shown in the SEM images. Carbon was deposited in (b) both barrels or (c) one barrel. (Panels a and b, reprinted with permission from McKelvey, K. et al., Fabrication, characterization, and functionalization of dual carbon electrodes as probes for scanning electrochemical microscopy (SECM), *Anal. Chem.*, Vol. 85, 2013: pp. 7519–7526. Copyright 2011, American Chemical Society; Panel c, Takahashi, Y. et al., Multifunctional nanoprobes for nanoscale chemical imaging and localized chemical delivery at surfaces and interfaces, *Angew. Chem., Int. Ed.*, 2011. Vol. 50. pp. 9638–9642. Copyright 2011, Wiley-VCH Verlag GmbH & Co. KGaA. Reproduced with permission; Panel d, reprinted with permission from Takahashi, Y. et al., *Proc. Natl. Acad. Sci. USA*, Vol. 109, 2012: pp. 11540–11545.)

Chemical vapor deposition (CVD) was also employed for the preparation of carbon nanotips. Mirkin and coworkers deposited a CVD carbon layer from methane inside a quartz nanopipette with a tip radius of down to 50 nm (Figure 1.23a) [69]. The thickness and distribution of the carbon layer were controlled by the CVD time, which was normally 3 h to completely fill the pipette orifice with carbon. Minor variations in the deposition process gave the carbon-filled electrode that was either flat or recessed. Interestingly, the cavity of a recessed carbon nanotip was filled by electrochemically depositing platinum to fabricate a sharp and highly electroactive Pt nanotip for intracellular electrochemical measurements. In addition, the CVD-grown carbon nanotips were recently modified to serve as electrochemical nanosamplers [142] and resistive-pulse sensors [143]. Baker and coworkers prepared carbon nanotips by the CVD deposition of parylene on the outer wall of a pulled quartz nanorod or nanopipette [144]. The thin parylene film was pyrolyzed under a nitrogen atmosphere to produce conductive carbon. The carbon-coated nanorod was insulated with CVD parylene and milled by FIB technology to prepare a carbon-ring nanotip with an inner radius of ~350 nm and an outer radius of ~800 nm (Figure 1.23b) for SECM imaging. A carbon ring/nanopore tip was also prepared by using a nanopipette as a template (Figure 1.23c) for SECM–SICM imaging (Section 1.5.3). For instance, a carbon nanoring with an outer radius of 485 nm and an inner radius of 295 nm was formed around a quartz nanopipette with an inner radius of 220 nm.

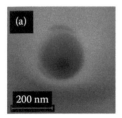 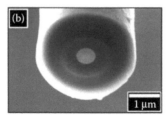 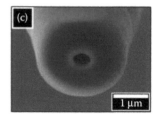

FIGURE 1.23 (a) SEM image of a CVD-carbon-filled quartz nanopipette. (With kind permission from Springer Science + Business Media, Hu, K. et al., Platinized carbon nanoelectrodes as potentiometric and amperometric SECM probes, *J. Solid State Electrochem.*, Vol. 17, 2013: pp. 2971–2977.) SEM images of (b) carbon nanoring and (c) carbon nanoring/nanopore tips. (Thakar, R. et al., Multifunctional carbon nanoelectrodes fabricated by focused ion beam milling, *Analyst*, Vol. 138, 2013: pp. 5973–5982. With permission from the Royal Society of Chemistry.)

1.4.2 NANOTIP CHARACTERIZATION

The shape and size of a nanotip must be accurately determined for nanoscale SECM, especially when data is quantitatively analyzed or spatial resolution is quantitatively assessed. It should be emphasized that, without the knowledge of tip shape, tip size can be readily underestimated or overestimated from a limiting current in the bulk solution, which can be conveniently measured by amperometry and voltammetry. The shape and shape of a nanotip can be much more reliably characterized by SECM. An important advantage of the SECM-based characterization is that the effective shape and size of a nanotip can be determined in solution. This *in situ* SECM characterization contrasts to *ex situ* SEM, FIB, and TEM characterization. Recently, AFM was introduced as a unique method for both *in situ* and *ex situ* tip characterization.

1.4.2.1 Effects of Tip Size and Shape on Bulk Limiting Current

A limiting current at an SECM tip in the bulk solution substantially depends on the shape and size of both electrochemically active and inactive portions of a tip, thereby prohibiting the determination of these four parameters only from a limiting current. For instance, a limiting current at an inlaid disk tip with any RG in the bulk solution, $i_{T,\infty}$, is given by Reference [145]

$$i_{T,\infty} = 4xnFDca \qquad (1.6)$$

with

$$x = 1 + 0.639\left[1 - \frac{2}{\pi}\arccos\left(\frac{1}{RG}\right)\right] - 0.186\left\{1 - \left[\frac{2}{\pi}\arccos\left(\frac{1}{RG}\right)\right]^2\right\} \qquad (1.7)$$

where n is the number of transferred charge per reactant molecule, F is Faraday constant, and c is the bulk concentration of the reactant. The x value varies from 1.3

with $RG = 1.1$ to ~1.0 with $RG > 10$. This ~30% variation of x value is larger than an experimental error in the measurement of a limiting current. Thus, both inner and outer radii of an inlaid disk tip cannot be accurately determined from a limiting current by using Equations 1.6 and 1.7. In addition to a and RG values, a limiting current at a recessed disk tip strongly depends on recession depth [81]. Equation 1.6 is valid for a recessed disk tip with $RG = 10$ when x is given by Reference [146]

$$x = \frac{1}{1.0354 + 1.261H + 0.01151\ln H} \qquad (1.8)$$

where H is the aspect ratio of the recession depth with respect to the disk radius. Equation 1.8 fits simulated values within a 0.4% error at $0.01 \leq H \leq 5$. The x value decreases as H increases, that is, from $x = 1$ with $H = 0$ to $x = 0.880$ with $H = 0.1$ and $x = 0.1358$ with $H = 5.0$. Thus, both disk radius and recession depth cannot be determined from a limiting current by using Equations 1.6 and 1.8. Moreover, a disk radius is underestimated from a limiting current at a recessed tip by using Equations 1.6 and 1.7 with the assumption of inlaid disk geometry. By contrast, this assumption results in the overestimation of a radius for a conical tip, which gives a higher limiting current than the inlaid disk tip with the same radius [147]. Quantitatively, a limiting current at a conical tip with a base radius of a is given by Equation 1.6, where x varies with the aspect ratio of the cone height with respect to the base radius, H, as given by Reference [148]

$$x = 1 + 0.30661H^{1.14466} \qquad (1.9)$$

This equation gives an error of <1% with $0 < H < 3$ and $RG = 100$. The x value increases as H increases, for example, $x = 2.1$ with $H = 3$ in Equation 1.9 and $x = 4.5$ with $H = 10$ from the numerical simulation [148]. Again, both base radius and cone height cannot be determined from a limiting current at a conical tip by using Equations 1.6 and 1.9. Noticeably, the analysis of a limiting current can be more reliable when the geometry and size of a nanotip are characterized additionally by SEM. The insulating portion of a nanotip including a glass nanopipette is less severely charged during SEM characterization when a thin metal film is carefully coated on the insulating surface. In fact, a tip size of down to ~10 nm can be estimated for a glass nanopipette with knowledge of a film thickness [149]. Such a metal-coated tip, however, is not useful for SECM measurement. Moreover, the *ex situ* size and geometry can be different from *in situ* ones, for instance, owing to tip damage. In addition, SEM is not applicable to the characterization of a deeply recessed nanotip or the nanopipette tip based on the ITIES, which can be formed only in liquids.

1.4.2.2 *In Situ* SECM Characterization

The SECM theories were developed for various tip shapes to enable the *in situ* quantitative characterization of a nanotip by approach curve measurement in solution. The fit of an experimental approach curve with a theoretical curve provides quantitative information about the shape and size of an SECM tip as originally proposed in early 1990s [6]. Since then, SECM theories have been developed for approach curves with disk [145,150], ring [151], conical [152–153], hemispherical [152,154], sphere cap [155–156], and recessed [81] tips. Presently, theoretical approach curves can be readily obtained for a tip with almost any geometry by numerically solving 2D or 3D diffusional problems. The numerical simulation is often done by using the powerful commercial software based on the finite element method, COMSOL Multiphysics. In addition, simulation results for typical tip geometries such as inlaid disk have been summarized as empirical equations for convenient use. For instance, a negative feedback approach curve at an insulating substrate is given for an inlaid disk tip with any RG by Reference [145]

$$\frac{i}{i_{T,\infty}}$$

$$= \frac{(2.08/RG^{0.358})(L - (0.145/RG)) + 1.585}{(2.08/RG^{0.358})(L + 0.0023RG) + 1.57 + (\ln RG/L) + (2/\pi RG)\ln(1 + (\pi RG/2L))}$$

$$(1.10)$$

where $L = d/a$. A negative approach curve with an inlaid disk tip is very sensitive to tip RG and is useful for its determination. Figure 1.24a shows a negative approach curve with the ITIES tip supported by a mechanically polished nanopipette as compared with theoretical curves with various RG values in Equation 1.10 [98]. Tip current was based on the transfer of tetraethylammonium from an outer aqueous solution into a

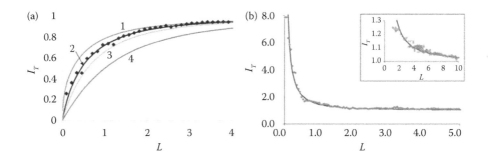

FIGURE 1.24 (a) Experimental (symbols) approach curve for tetraethylammonium transfer from water to an 8-nm-radius DCE-filled pipette approaching a solid substrate. Theoretical curves (solid lines) represent Equation 1.10 with $a = 8.1$ nm, and $RG=$ (a) 1.1, (2) 1.9, (3) 2.5, (4) and 10. (Reprinted with permission from Elsamadisi, P. et al., Polished nanopipets: New probes for high-resolution scanning electrochemical microscopy, *Anal. Chem.*, Vol. 83, 2011: pp. 671–673. Copyright 2011, American Chemical Society.) (b) Experimental approach curve (symbol) obtained with a 46-nm-radius Pt tip at an Au substrate in the aqueous solution containing 1 mM FcCH$_2$OH and 0.2 M NaCl. The theoretical curve (solid line) represents the diffusion-controlled positive feedback curve. The inset shows experimental and theoretical approach curves for a 13-nm-radius Pt tip. (Reprinted with permission from Sun, P. and M. V. Mirkin, Kinetics of electron-transfer reactions at nanoelectrodes, *Anal. Chem.*, Vol. 78, 2006: pp. 6526–6534. Copyright 2006, American Chemical Society.)

DCE-filled nanopipette with an inner radius of 8.1 nm. The best fit was obtained with $RG = 1.9$, while the experimental curve significantly deviates from theoretical curves with larger or smaller RG values of 1.1 or 2.5, respectively. This analysis also gave an extremely short distance of ~0.8 nm at tip–substrate contact, which confirmed that the orifice of the polished nanopipette is very flat. The flatness of a tip can be checked more precisely by a positive approach curve, which is more sensitive to the tip–substrate distance. A positive approach curve with an inlaid disk tip with any RG is empirically given by Reference [150]

$$\frac{i}{i_{T,\infty}} = y + \frac{\pi}{4x\arctan(L)} + \left[1 - y\frac{1}{2x}\right]\frac{2}{x}\arctan(L) \qquad (1.11)$$

with

$$y = \ln 2 + \ln 2\left[1 - \frac{2}{\pi}\arccos\left(\frac{1}{RG}\right)\right] - \ln 2\left\{1 - \left[\frac{2}{\pi}\arccos\left(\frac{1}{RG}\right)\right]^2\right\} \qquad (1.12)$$

Figure 1.24b shows that the current based on FcMeOH oxidation at a 46-nm-radius Pt tip is enhanced by a factor of 8 as the tip approaches to ~5 nm distance from a flat Au substrate [83]. This close distance supports the flat geometry of the nanotip without significant recession or protrusion, which results in a lower positive feedback response (see below). In contrast, ~13-nm-radius Pt tip gave much lower positive feedback current to contact the substrate with an apparent tip–substrate distance of ~26 nm (the inset of Figure 1.24b) despite the good fit of the experimental curve with the theoretical curve.

Recessed or protruded nanotips are useful for certain applications, although they give smaller feedback responses than flat nanotips. For instance, recessed nanotips were used for single-molecule detection to trap a molecule in the cavity (Figure 1.13a) [80] as well as for intermittent-contact imaging to protect a tip from the substrate (Figure 1.6) [32]. Figure 1.25a shows a positive feedback curve at the Pt nanotip that was recessed by electrochemical etching [81]. The current at the recessed tip increased only up to twice of the limiting current in the bulk solution. The experimental curve fitted well with a theoretical curve to yield $a = 84.3$ nm and $H = 0.31$ (i.e., ~26 nm recession depth). Similarly, a small positive feedback response was observed with a protruded Pt nanotip (Figure 1.25b) [157]. This tip was protruded by electrochemically depositing mercury on the Pt surface. The experimental curve was much less positive than expected for an inlaid disk tip. A good fit was obtained with a theoretical curve for a hemispherical tip to yield a tip radius of 67 nm and an extremely short tip–substrate distance at their contact. Noticeably, current at the hemispherical tip did not level off at the tip–substrate contact, which contrasts to the behavior of the recessed tip (Figure 1.25a). The tip current at the recessed tip leveled off because the insulating sheath of the tip contacted the conductive substrate as assumed for Equation 1.4. Thus, recessed and protruded tip geometries can be discriminated by using a conductive substrate despite their similarly low positive feedback responses. Alternatively, an insulating substrate can be used to discriminate

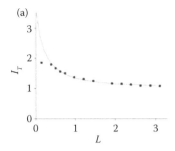
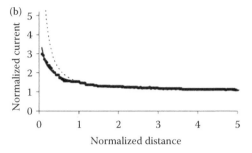

FIGURE 1.25 (a) Experimental approach curve (squares) as obtained with an etched Pt tip at an Au film substrate in 1 mM FcCH$_2$OH and 0.2 M KNO$_3$. Theoretical fit (solid line) gave $a = 84.3$ nm and $H = 0.31$. (Reprinted with permission from Sun, P. and M. V. Mirkin, Scanning electrochemical microscopy with slightly recessed nanotips, *Anal. Chem.*, Vol. 79, 2007: pp. 5809–5816. Copyright 2007, American Chemical Society.) (b) Experimental approach curve (symbols) obtained with a Hg-capped Pt nanotip at an Au substrate in 2 mM Ru(NH$_3$)$_6$Cl$_3$. Theoretical curve (solid line) was calculated for a 67-nm-radius hemispherical tip. Dashed curve is the theory for a disk shaped tip. (Velmurugan, J. and M. V. Mirkin, Fabrication of nanoelectrodes and metal clusters by electrodeposition, *ChemPhysChem.*, 2010. Vol. 11. pp. 3011–3017. Copyright 2010, Wiley-VCH Verlag GmbH & Co. KGaA. Reproduced with permission.)

recessed and protruded tip geometries from approach curves [137], which give zero and non-zero tip currents, respectively, at the zero tip–substrate distance.

1.4.2.3 AFM Characterization in Air and in Solution

AFM was recently introduced as a powerful tool to provide detailed information about the geometry and surface reactivity of a metal nanotip both in air and in solution [158]. For instance, the non-contact AFM image of a Pt nanotip in air clearly demonstrated its recession (Figure 1.26a). This image gave a tip radius of ~52 nm and a recession depth of ~42 nm. Equation 1.8 predicts that a limiting current at the recessed tip with $H = $ ~0.8 is nearly a half of that at the inlaid disk electrode with the same radius [81]. This prediction was confirmed experimentally, where a limiting current of ~9 pA as measured in 1.2 mM FcCH$_2$OH agreed well with a value of 9.4 pA calculated from Equation 1.8. Importantly, the recessed geometry is unnoticeable from the corresponding steady-state voltammogram with a well-defined sigmoidal shape. Without considering the recessed geometry, a tip radius is significantly underestimated to be ~20 nm from the experimental limiting current by assuming inlaid disk geometry. More recently, *ex situ* AFM imaging was also used to characterize a recessed Ag nanotip with an effective radius of ~50 nm and a recession depth of ~6 nm [126].

AFM imaging can be performed in solution to monitor a change in tip geometry by electrochemical deposition or dissolution. Figure 1.26b shows the *in situ* AFM image of a Pt nanotip that was taken immediately after platinization in solution [68]. Conveniently, the degree of platinization can be checked by AFM topography imaging to be adjusted for specific applications. Importantly, non-contact AFM imaging

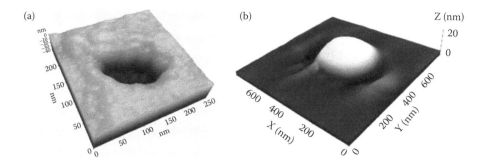

FIGURE 1.26 (a) Non-contact AFM image of a recessed Pt nanotip in air. (Reprinted with permission from Nogala, W., J. Velmurugan, and M. V. Mirkin, Atomic force microscopy of electrochemical nanoelectrodes, *Anal. Chem.*, Vol. 84, 2012: pp. 5192–5197. Copyright 2012, American Chemical Society.) (b) *In-situ* AFM image of a 115-nm-radius Pt tip after platinization. (Reprinted with permission from Wang, Y. X. et al., Nanoelectrodes for determination of reactive oxygen and nitrogen species inside murine macrophages, *Proc. Natl. Acad. Sci. USA*, Vol. 109, 2012: pp. 11534–11539.)

does not damage the tip surface, thereby maintaining its electroactivity. Moreover, tip topography can be characterized by *in situ* AFM without interference from contamination of the tip surface, which can be caused during cleaning and drying of the tip for *ex situ* characterization. Recently, the unexpected electrochemical dissolution of a Pt nanotip was revealed by *in situ* AFM imaging [159].

1.4.3 DAMAGE AND CONTAMINATION OF NANOTIP

The nanoscale damage and organic contamination of a nanotip can be serious problems, which have been unnoticed until recently. For instance, a Lewis paper in 1990 reported the voltammetry of a glass-sealed Pt nanoelectrode, where the Pt tip looked completely covered with glass as observed by SEM [160]. The origin of tip recession was never addressed or understood. A recent study demonstrated that micrometer- or nanometer-sized Pt tips can be damaged at the nanoscale by electrostatic discharge (ESD) [124]. The recessed geometry of ESD-damaged tips explains at least partially why Pt nanoelectrodes often lose a current response [88] or give a low current in cyclic voltammetry [32,35] and in SECM approach curves [158]. Importantly, the nanoscale ESD damage of a tip by an operator can be prevented by grounding the tip and the operator. In addition, relatively high humidity (>30% at 20°C) must be maintained to prevent ESD damage [130]. Figure 1.27a shows severe ESD damage on a glass-sealed Pt tip with a radius of 0.47 μm and a small *RG* of ~2 as prepared by using a CO_2-laser puller and an FIB instrument. No damage was seen in the *in situ* SEM image of this tip immediately after FIB milling (Figure 1.19a). Severe damage was made to the tip end when the contact wire of the submicrometer-sized Pt tip was touched by the bare hand of an operator without ESD protection. The operator did not notice such damage upon contact because the operator did not see, hear, or even feel ESD. The ESD-damaged

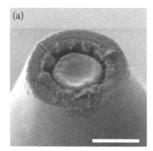
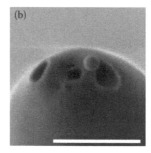

FIGURE 1.27 SEM images of ESD-damaged Pt tips. See the corresponding panels of Figure 1.19a and b for the SEM images of the same tips before ESD damage. Scale bars are 1 μm. (Reprinted with permission from Nioradze, N. et al., Origins of nanoscale damage to glass-sealed platinum electrodes with submicrometer and nanometer size, *Anal. Chem.*, Vol. 2013: pp. 6198–6202. Copyright 2013, American Chemical Society.)

Pt tip was recessed and its periphery was deformed. The severely ESD-damaged tip gave a poor SECM approach curve with a small feedback effect from an Au substrate as expected for a recessed tip. The experimental approach curve was too broad to fit with the theoretical approach curve based on the positive feedback effect. The recession of the damaged tip, however, was not obvious from its excellent steady-state cyclic voltammogram. More severe ESD damage was seen with ~55-nm-radius Pt tip (Figures 1.19b and 1.27b before and after ESD damage, respectively). The ESD-damaged Pt nanotip was completely buried into the glass sheath. The damaged nanotip gave a limiting current of only 1.0 pA–0.5 mM FcMeOH in 0.1 M KCl. When inlaid disk geometry is assumed, this current corresponds to a tip radius of 6.1 nm, which is ~9 times smaller than the radius of the Pt tip before ESD damage. This result indicates that the size of a Pt nanotip can be significantly underestimated from its limiting current when the electrode is unknowingly recessed owing to ESD damage.

A metal nanotip must be protected also from nanoscale electrochemical damage due to the transient flow of high current from a potentiostat [124]. Figure 1.28a shows the SEM image of ~0.49-μm-radius Pt tip after measurement of 38 cyclic voltammograms of 0.5 mM FcMeOH in 0.1 M KCl at a tip potential between −0.05 and 0.4 V. In this potential range, Pt is electrochemically stable. The SEM image, however, shows ~250-nm-depth recession of the Pt tip with adsorbates on the surface of the glass sheath. X-ray energy dispersive spectroscopy of the damaged tip confirmed that the adsorbates were Pt nanoparticles (Figure 1.28b), which were etched away from the Pt tip. Electrochemical etching of a Pt tip is due to the transient tip current driven by the saturation of a counter electrode amplifier upon the switching of the potentiostat from the internal dummy cell to the electrochemical cell (or vice versa) (Figure 1.28c). This transition was initiated by turning on relay switches for counter and reference electrodes. Since electromechanical relays bounced and did not settle down immediately, the counter-electrode amplifier transiently experienced an open loop configuration to be saturated. The resultant large output voltage of the saturated amplifier transiently drove high current from the Pt tip through the

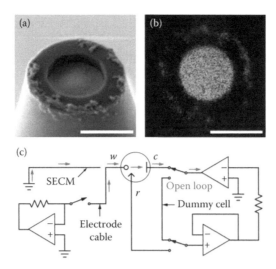

FIGURE 1.28 (a) SEM and (b) X-ray energy dispersive spectroscopic images of a Pt tip after measurement of the 38 cyclic voltammograms of 0.5 mM FcMeOH in 0.1 M KCl. Scale bars are 1 μm. In (b), orange dots indicate the presence of Pt. (c) Flow of transient current from a Pt tip to a saturated counter-electrode amplifier under an open-loop configuration at the beginning or end of voltammetric measurement. Arrows indicate current flow while w, r, and c represent working (tip), reference, and counter electrodes. (Reprinted with permission from Nioradze, N. et al., Origins of nanoscale damage to glass-sealed platinum electrodes with submicrometer and nanometer size, *Anal. Chem.*, Vol. 2013: pp. 6198–6202. Copyright 2013, American Chemical Society.)

electrolyte solution, thereby electrochemically etching the tip. The current was not only sufficiently high to etch the Pt tip but also low enough to be supplied capacitively from the working electrode (tip) cable, which was not connected to the potentiostat. Electrochemical damage to a Pt tip was almost completely suppressed by maintaining the internal connection of the potentiostat to the electrochemical cell between voltammetric measurements.

The contamination of nanotips with organic impurities in ultrapure water is the likely but experimentally unaddressed problem, which has been pointed out only theoretically. Scherson and Tolmachev argued that the oxygen reduction reaction at Pt nanoelectrodes forms hydrogen peroxide rather than water because the surface active sites are blocked by the organic contaminants that are irreversibly adsorbed from water [161]. They assumed the diffusion-limited adsorption of the contaminants to estimate that the surface of a 100-nm-diameter Pt electrode can be fully covered with organic impurities in 12 s even when their concentration corresponds to a total organic carbon of only 1 ppb (e.g., 13 nM for benzene), which is the lowest limit achievable by using commercial water purifiers on the market. This extremely fast coverage is due to the efficient radial diffusion of organic contaminants to the nanoscale Pt surface and also to their irreversible adsorption. Originally, Chen and Kucernak concluded from their experiments that the dominant production of hydrogen peroxide from the oxygen reduction reaction at a nanoelectrode modified with a single Pt nanoparticle was due to the efficient diffusional loss of hydrogen peroxide from the nanoelectrode surface before its reduction to water [162]. The organic contamination of a nanotip, if any, will complicate the interpretation and quantitative analysis of the results of nanoscale SECM measurements.

1.4.4 NANOPOSITIONING

Controlling the position of a nanotip with the accuracy and precision of nanometer and subnanometer scales is crucial to nanoscale SECM. Both vertical and lateral tip positions must be accurate and repeatable for constant-distance imaging. In addition, vertical tip position must be stable during constant-height imaging and nanogap experiments. Presently, a few commercial SECM instruments employ closed-loop piezoelectric positioners for tip nanopositioning (Table 1.1). A closed-loop piezoelectric positioner is equipped with a high-resolution sensor to monitor the position of the stage. The sensor response is used as a feedback signal to dynamically control the stage position, thereby minimizing an error in tip position due to creeping and hysteresis of the piezoelectric stage. The commercial SECM systems use strain gauge sensors for x, y, and z axes or a capacitive sensor only for the z axis. A piezoelectric stage with a strain gauge sensor is less expensive while a capacitive sensor provides superior resolution and repeatability of <0.1 nm, which also depends on the travel range of the piezoelectric stage and the type of the stage controller. A problem of a strain gauge sensor is its higher self-heating, which may cause the thermal drift of tip position (see below). Home-built SECM instruments for nanoscale experiments are typically based on closed-loop piezoelectric stages with capacitive sensors (e.g., Figure 1.6 [32]). Useful information about piezoelectric nanopositioning systems is available at the Web site of their manufacturer [163].

Nanoscale tip positioning is highly challenging under ambient conditions because of thermal drift. Thermal drift is a general problem in nanoscale imaging by various

TABLE 1.1
Commercial SECM Instruments with Closed-Loop Piezoelectric Positioners

Manufacturer	Model	Position Sensor[a]	Pt Tip Diameter (μm)
CH Instruments	CHI 920D[b]	Strain gauge	10 and 25
Sensolytics	High-Res option[c]	Strain gauge	0.2–0.5, 1, 5, 10,[d] 25,[d] 50,[d] and 100
HEKA	ElProScan[e]	Capacitive[f]	0.2–0.5, 0.5–1, 4.7,[g] 10,[d] and 25

[a] Information obtained by direct contact with the manufacturer.

[b] http://www.chinstruments.com/chi900.shtml.

[c] http://www.sensolytics.eu/en/products2/secm/secm-systems/170-03-00002.

[d] Also available for gold tips.

[e] http://www.heka.com/products/systems_elproscan.html.

[f] Only for the z axis.

[g] Only available for carbon fiber tips.

scanning probe techniques and causes vertical and lateral image distortions even when the probe–substrate distance is feedback controlled. For instance, thermal drift can be caused by the expansion and contraction of an SECM stage upon a slight change in the temperature of the surrounding environment [99]. The ~10 cm-height SECM stage made of a material with a typical coefficient of linear thermal expansion in the order of $10^{-5}/K$ expands (or contract) by 10 nm for an increase (or decrease) in temperature by 10 mK. Ambient temperature in a laboratory usually changes much more, especially, in the presence of an operator as a heat source. In fact, the significant drift of tip–substrate distance in the ambient environment was confirmed by monitoring the negative feedback current at ~0.5-μm-radius Pt or ITIES tip positioned within a feedback distance from an insulating substrate without any tip movement. Figure 1.29a shows the time profile of a current response at a 0.44-μm-radius Pt tip with a 0.22-μm-thick glass sheath when the tip was brought to the surface of a SiO_2/Si wafer. The tip was positioned at 0.11 μm from the substrate, where the tip–substrate distance was calculated from tip current by using the inverse function of Equation 1.10 (Figure 1.29b). As soon as the tip approach was stopped, the tip current gradually increased to nearly completely recover to $i_{T,\infty}$ within ~10 min, where the negative feedback effect became almost negligible. This result indicates that the tip–substrate gap became wider without moving the z-axis piezoelectric positioner owing to thermal drift. A total drift of up to 1 μm is too large to be ascribed to the hysteresis or creeping of a piezoelectric positioner, which was feedback controlled by a capacitive position sensor. Such a time profile was

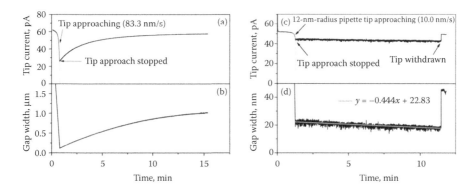

FIGURE 1.29 Time profiles of (a) the tip current based on the oxidation of 0.5 mM FcMeOH in 0.2 M NaCl as obtained using a 0.44-μm-radius Pt tip with $RG = 1.5$ without an isothermal chamber and (b) the corresponding width of the tip–substrate gap. Time profiles of (c) the tip current based on the transfer of 10 mM tetraethylammonium in 0.3 M KCl as obtained by using ~12-nm-radius nanopipette filled with DCE in an isothermal chamber and (d) the corresponding tip–substrate gap width. Chamber temperature drifted at −0.2 mK/min during the current measurement. Theoretical curves in (a) and (c) were calculated by using the inverse function of Equation 1.10. (Reprinted with permission from Kim, J. et al., Stabilizing nanometer scale tip-to-substrate gaps in scanning electrochemical microscopy using an isothermal chamber for thermal drift suppression, *Anal. Chem.*, Vol. 84, 2012: pp. 3489–3492. Copyright 2012, American Chemical Society.)

measured on different days to find that the gap became either wider or narrower at drift rates in a wide range of 5–150 nm/min.

A new isothermal chamber was developed to prevent thermal drift [99], which was only correctable as practiced for AFM or unavoidable unless a cryostat condition or fast scanning was employed. Air temperature in the isothermal chamber changed only at <0.2 mK/min to remarkably and reproducibly slow down the drift of tip position to <0.4 nm/min. The subnanometer stability of a tip–substrate nanogap in the isothermal chamber was confirmed by using a nanopipette-supported ITIES tip. An $i_{T,\infty}$ value of 52 pA was based on the transfer of tetraethylammonium across the ITIES formed at the ~12-nm-radius tip of a DCE-filled nanopipette (Figure 1.29c). The tip approached to the SiO_2/Si wafer surface until the tip current dropped to ~85% of $i_{T,\infty}$, which corresponds to a tip–substrate distance of 22.3 nm (Figure 1.29d). After tip approach was stopped, the tip current decreased only by ~3 pA for 10 min, which corresponds to a decrease of 4.6 nm in the tip–substrate distance. A drift rate of −0.44 nm/min was obtained from the linear fit of gap width versus time in Figure 1.29d. Remarkably, this drift rate is 11–340 times lower than that without the chamber and is comparable to or better than that of 0.6–6 nm/min as reported for electrochemical STM/spectroscopy [164]. Eventually, the stability of a tip–substrate gap in the isothermal chamber was limited by the subnanometer-scale fluctuation of the gap, which resulted in noisier tip current at feedback distances than in the bulk solution (Figure 1.29c). The corresponding distance fluctuation with respect to the best linear fit in Figure 1.29d gave a standard deviation of ±0.9 nm. This fluctuation was mainly ascribed to the dynamic instability of the piezoelectric positioner. Noticeably, thermal drift was not a problem when a nanotip contacted a substrate for single-molecule detection (Figure 1.12) [78].

1.5 COMBINED TECHNIQUES FOR NANOSCALE IMAGING

The unique capability of SECM to image chemical reactivity and topography at interfaces renders this powerful technique complementary to and even combinable with other scanning probe microscopy techniques. Specifically, SECM imaging can be combined with topography imaging by AFM or more recently by SICM to obtain a reactivity image in the constant-distance mode. Alternatively, an optical image can be obtained by scanning optical microscopy while SECM allows for constant-distance topography imaging. Technologically, multifunctional nanotips are required to enable multidimensional imaging by combined SECM techniques [165]. Various nanofabrication techniques such as FIB milling were introduced for the development of multifunctional nanotips to make broad impact on nanotip fabrication in general as discussed in Section 1.4.1.

1.5.1 MULTIFUNCTIONAL NANOTIPS

Multifunctional nanotips were developed for multidimensional nanoscale imaging by combined SECM techniques. Primarily, a multifunctional nanotip serves as a nanoelectrode to provide reactivity information about a target interface while the second function of the nanotip is to provide information about the tip–substrate distance for constant-distance topography imaging. For instance, the fabrication of an AFM cantilever from a nanoelectrode gives a dual functional tip (Figure 1.30a,b) to enable SECM reactivity imaging and AFM topography imaging [61]. Originally, an SECM–AFM tip was developed to electrochemically drive a surface reaction, for example, crystal dissolution, and investigate the reaction-induced change of surface topography in the AFM mode [166]. By contrast, a tip–substrate distance can be monitored electrochemically for topography imaging when the other function of the tip is to obtain an optical image. For instance, the SECM tip based on an optical fiber was developed to serve not only as a ring ultramicroelectrode but also as a light source for simultaneous optical imaging (Figure 1.30c) [167]. Tip current was monitored for constant-distance topographic imaging during optical imaging [30]. The optical-fiber tip was also used to obtain electrochemical and optical images in the constant-distance mode while a constant tip–substrate distance was maintained by measuring shear force between the tip and the substrate in the hopping mode [30]. Later, optical-fiber-based tips were miniaturized to nanometer size to improve the spatial resolution of SECM combined with scanning optical [168] or near-filed microscopy [169].

Elaborate multifunctional nanotips have been developed by employing various nanofabrication techniques. The nanofabrication approach was pioneered by Kranz and coworkers, who introduced thin-film deposition and FIB-milling technologies [170] to fabricate a square-frame-shaped nanoelectrode beneath the tip end of an AFM cantilever (Figure 1.31a) [171]. Such an SECM–AFM tip has been successfully used for simultaneous reactivity and topography imaging in both contact [171–172] and tapping [173] modes. A frame-shaped electrode was also modified with enzymes such as glucose oxidase to serve as an integrated nanoscale biosensor, thereby enabling the selective imaging of glucose transport through nanoporous membranes [174]. More recently, the SECM–AFM nanotip based on a triangular frame electrode was nanofabricated to detect the topography and reactivity of individual single-wall carbon nanotubes in their network [175]. The spatial resolution of these square- and triangular-frame-shaped electrodes,

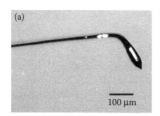
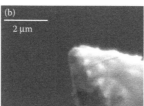
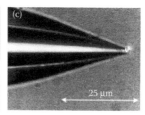

FIGURE 1.30 (a) Optical and (b) SEM images of an SECM–AFM tip. (Reprinted with permission from Macpherson, J. V. and P. R. Unwin, Combined scanning electrochemical-atomic force microscopy, *Anal. Chem.*, Vol. 72, 2000: pp. 276–285. Copyright 2000, American Chemical Society.) (c) Optical microscopy image of an optical-fiber-based ring ultramicroelectrode with coupling of Ar ion laser light at 514 nm. (Reprinted with permission from Lee, Y. and A. J. Bard, Fabrication and characterization of probes for combined scanning electrochemical/optical microscopy experiments, *Anal. Chem.*, Vol. 74, 2002: pp. 3626–3633. Copyright 2002, American Chemical Society.)

however, has been limited by their edge lengths of 300–800 nm. Moreover, the approach of a frame electrode to a substrate is limited by the 250–400-nm-height protrusion of the AFM tip from the electrode surface. These limitations of the frame-shaped design can be overcome by integrating a nanoelectrode at the very tip of an AFM cantilever [176–180]. Alternatively, a single-wall carbon nanotube was attached to the tip of an AFM cantilever as the template of a nanoelectrode (Figure 1.31b) [181]. The nanotube was coated with gold, insulated with polymer and silicon nitride layers, and milled by FIB to expose a tip with ~100 nm diameter. Such a high-aspect-ratio SECM–AFM tip was also fabricated by employing an Ag_2Ga alloy needle as a template [182].

Noticeably, the nanofabrication approach was also successful for the fabrication of the SECM–SICM nanotips based on heat-pulled nanopipettes [144,183,184]. For instance, a heat-pulled nanopipette was coated with gold, insulated by the atomic layer deposition of aluminum oxide, and milled by FIB technology to expose a 100-nm-diameter nanopore next to a gold nanoelectrode with an effective radius of 294 nm (Figure 1.31c) [183]. Similarly, an SECM-SICM nanotip was

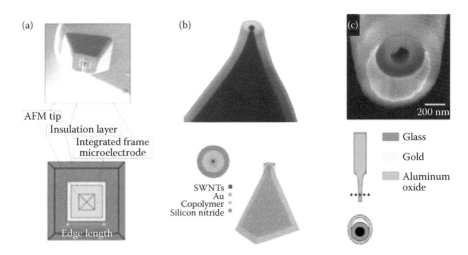

FIGURE 1.31 (a) SEM image and schematic top view of the frame electrode integrated into an AFM cantilever. (Kueng, A. et al., AFM-tip-integrated amperometric microbiosensors: High-resolution imaging of membrane transport, *Angew. Chem., Int. Ed.*, 2005. Vol. 44. pp. 3419–3422. Copyright 2005, Wiley-VCH Verlag GmbH & Co. KGaA. Reproduced with permission.) (b) FIB image and schematic view of an SECM–AFM tip with a single-walled carbon nanotube as a template. (Reprinted with permission from Burt, D. P. et al., Nanowire probes for high resolution combined scanning electrochemical microscopy-atomic force microscopy, *Nano Lett.*, Vol. 5, 2005: pp. 639–643. Copyright 2005, American Chemical Society.) (c) SEM image and schematic view of an SECM-SICM nanotip. The tip was FIB-milled at the dotted line to expose a nanopore and a gold nanoelectrode. (Reprinted with permission from Comstock, D. J. et al., Integrated ultramicroelectrode-nanopipet probe for concurrent scanning electrochemical microscopy and scanning ion conductance microscopy, *Anal. Chem.*, Vol. 82, 2010: pp. 1270–1276. Copyright 2010, American Chemical Society.)

developed by coating pyrolytic carbon on the outer wall of a nanopipette as shown in Figure 1.23c [144].

1.5.2 Atomic Force Microscopy

The combination of SECM with AFM has been successful for simultaneous reactivity and topography imaging at the nanoscale [185–186]. A high spatial resolution of ~70 nm was demonstrated for electrochemical imaging by SECM–AFM using a metal-coated AFM cantilever [62]. In this approach (Figure 1.32a,b), the tip end of a Pt-coated AFM cantilever was immersed in a thin aqueous layer to serve as a nanoelectrode without insulation during simultaneous topography imaging by AFM [187]. A nanoporous polycarbonate membrane (100-nm-diameter pore size) was hydrated with an electrolyte solution containing a redox-active probe molecule

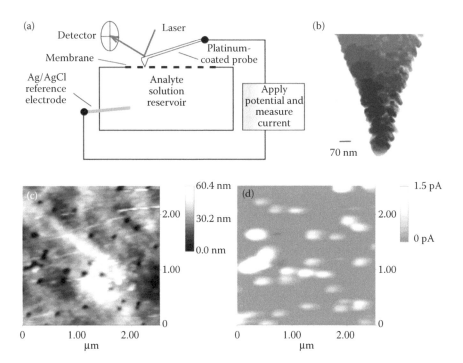

FIGURE 1.32 (a) Schematic representation of the experimental setup for SECM–AFM imaging of wetted nanopores. (Reprinted with permission from Gardner, C. E. and J. V. Macpherson, Atomic force microscopy probes go electrochemical, *Anal. Chem.*, Vol. 74, 2002: pp. 576A–584A. Copyright 2002, American Chemical Society.) (b) TEM image of a typical Pt-coated AFM cantilever. Simultaneous (c) height and (d) current images of a wetted track etched polycarbonate membrane. The tip was biased at +1.0 V versus Ag/AgCl for diffusion-limited oxidation of 10 mM $IrCl_6^{3-}$ in 0.5 M KNO_3. (Reprinted with permission from Macpherson, J. V. et al., Electrochemical imaging of diffusion through single nanoscale pores, *Anal. Chem.*, Vol. 74, 2002: pp. 1841–1848. Copyright 2002, American Chemical Society.)

such as $IrCl_6^{3-}$ or $Fe(phenanthroline)_3^{2+}$ to function as a model membrane system. The individual nanopores of the track-etched membrane was imaged by AFM (Figure 1.32c) while the diffusional transport of $IrCl_6^{3-}$ through the nanopores was imaged electrochemically (Figure 1.32d) [62]. The analysis of the corresponding current image based on the diffusion-limited oxidation of the redox molecule at the tip indicated that the solution was largely confined to pores in the membrane. In fact, the diffusion of 10 mM $IrCl_6^{3-}$ through the long pores (10 μm) resulted in a very low tip current of <1.5 pA over the nanopores. This tip current is much lower than tip current as obtained using a 30-nm-thick silicon nanopore membrane (Figure 1.8c). Noticeably, metallic AFM cantilevers were also insulated to enable nanoscale SECM–AFM imaging of the substrate that was completely soaked in solution. A Pt-coated AFM cantilever was insulated by an electrophoretic paint to expose only the tip end as a conical nanoelectrode with a radius of ~300 nm [188]. Analogously, metal wires were etched, bent, and insulated by an electrophoretic paint to serve as a cantilever for AFM instruments (Figure 1.30b,c) [61,189]. The SECM–AFM nanotip based on an insulated metallic cantilever was also employed for nanoscale electrochemical lithography [190].

The highest spatial resolution of ~9 nm for electrochemical reactivity imaging by SECM–AFM was achieved by employing the SiO_2-coated Pt nanoelectrode formed at the tip of an AFM cantilever [191]. The TEM image of the SECM–AFM nanotip (Figure 1.33a,b) shows the Pt tip (zone 3) exposed from the insulating SiO_2 layer (zone 1) by etching in buffered HF, which also resulted in formation of a gap (zone 2) between the SiO_2 insulation and the Pt tip. The SECM–AFM nanotip was used for topography and reactivity imaging of the 2 μm-wide Pt lines deposited on the Si_3N_4 surface (Figure 1.33c,d, respectively). First, the topography image was obtained to confirm a thickness of ~100 nm and an edge-edge separation of ~100 nm for the Pt lines. Topography imaging was followed by constant-distance electrochemical imaging, where the tip was scanned at a distance of ~30 nm from the substrate surface. The reactivity image was obtained by using $Ru(NH_3)_6^{3+}$ as a redox mediator to amperometrically observe positive and negative feedback effects from the Pt and Si_3N_4 surfaces, respectively. The imperfection of the left Pt line in the topography image (dotted line) is due to contaminations, which were confirmed by less positive feedback in the reactivity image. A space between the Pt lines in the electrochemical image is smaller than that in the topography image. This discrepancy was ascribed to a positive feedback effect from the Pt sidewall when the tip was positioned over the Si_3N_4 surface. A high spatial resolution of 9 nm was determined from the SECM image, where a distance of 9 nm was observed between 5% and 95% of the electrochemical current step when the tip moved from above the Pt line to above the Si_3N_4 surface (Figure 1.32e). This spatial resolution is apparently comparable to the radius of curvature at the tip end, r, and is significantly better than expected from the whole tip size as defined by the base radius, a, and height, h, of the conical tip (Figure 1.32a), which determine the dimension of a diffusion layer around a conical tip (Equation 1.9 [153]). The SiO_2-coated Pt nanotip was also applied for the high-resolution SECM–AFM imaging of *Deinococcus radiodurans* [192] and highly oriented pyrolytic graphite [177].

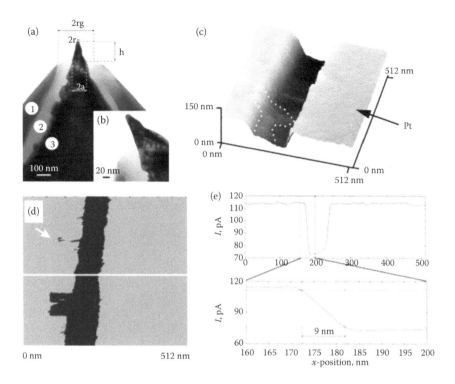

FIGURE 1.33 (a) and (b) TEM images of a SiO$_2$-coated Pt nanotip at an AFM cantilever. (c) Topography and (d) electrochemical images of Pt lines on the Si$_3$N$_4$ surface in 5 mM Ru(NH$_3$)$_6$Cl$_3$ and 1 M KCl. The dotted line in (c) indicates dirt over the edge of the left Pt line, whereas small dirt is indicated by the arrow in (d). (e) Profile of the electrochemical image along the horizontal line in (d). (Reprinted with permission from Gullo, M. R. et al., Characterization of microfabricated probes for combined atomic force and high-resolution scanning electrochemical microscopy, *Anal. Chem.*, Vol. 78, 2006: pp. 5436–5442. Copyright 2006, American Chemical Society.)

More recently, a conical Pt nanoelectrode with a base radius of ~200 nm was fabricated at the tip of an AFM cantilever to enable the topography and reactivity imaging of the graphene and graphite flakes exfoliated from highly oriented pyrolytic graphite [193]. Interestingly, this SECM–AFM study revealed the heterogeneous electroactivity of the graphitic surfaces that is not correlated to their topography. The heterogeneity was ascribed to the surface adsorption of chemical contaminants or intrinsic impurities. For instance, the topographic image of a graphite flak (Figure 1.34a) shows a uniform height of ~20 nm with a variation of <1 nm across the central region, where a low spot in electroactivity was seen in the corresponding electrochemical image (Figure 1.34b). By contrast, relatively similar electroactivity was observed when the nanoprobe was scanned over the monolayer graphene (~0.4 nm height) adjacent to the multiple layers of graphene (Figure 1.34c,d for topographic and electrochemical images, respectively). The cross sections of these images (Figure 1.34e) clearly demonstrate that the electroactivity of graphene layers

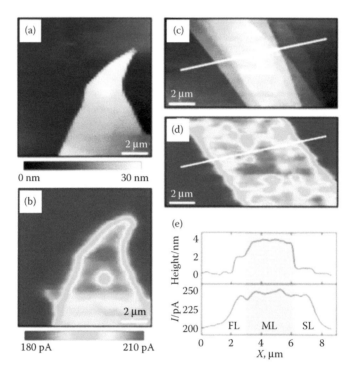

FIGURE 1.34 Feedback mode SECM–AFM images of exfoliated graphene/graphite flakes immersed in 1 mM FcMeOH/0.1 M KNO$_3$ solution. Topography is shown in (a) and (c) and the corresponding electrochemical scans are depicted in (b) and (d), respectively. (e) Line scan profiles of single-layer (SL), multilayer (ML), and few-layer (FL) regions of parts (c) and (d). Tip potential at 0.3 V versus Ag, line scan frequency 0.5 Hz, lift height 150 nm, bulk tip current typically ~200 pA. (Reprinted with permission from Wain, A. J., A. J. Pollard, and C. Richter, High-resolution electrochemical and topographical imaging using batch-fabricated cantilever probes, *Anal. Chem.*, Vol. 86, 2014: pp. 5143–5149. Copyright 2014, American Chemical Society.)

is independent of their thickness. This result apparently contradicts to the result of the recent study of monolayer and multilayer graphene by scanning electrochemical cell microscopy, where the electroactivity varied with graphene thickness [194]. Unfortunately, it was not determinable in this SECM–AFM study whether the apparently uniform electroactivity was limited simply by mediator diffusion between the cantilever tip and the graphene substrate or kinetically by mediator regeneration at the graphene substrate. Discrimination between diffusion and kinetic limitations requires the biasing of graphene samples and the quantitative analysis of tip current.

The frame-shaped electrode nanofabricated into an AFM cantilever has well-defined shape and size to enable quantitative image analysis by using the boundary element method [195]. The quantitative analysis was made on the SECM image of gold islands (20 × 20 µm) on the surface of an insulating substrate (Figure 1.35a). The size of the islands was more than 10 times larger than the integrated frame electrode (e.g., Figure 1.31a). Thus, the current based on Fe(CN)$_6^{4-}$ oxidation at the

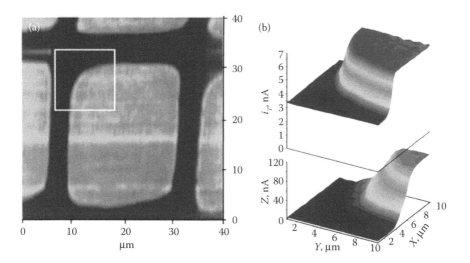

FIGURE 1.35 (a) SECM feedback image obtained with an integrated SECM–AFM electrode. (b) Simulated SECM feedback current (top) and the corresponding topography (bottom) in the area marked white in (a). (Reprinted with permission from Sklyar, O. et al., Numerical simulation of scanning electrochemical microscopy experiments with frame-shaped integrated atomic force microscopy-SECM probes using the boundary element method, *Anal. Chem.*, Vol. 77, 2005: pp. 764–771. Copyright 2005, American Chemical Society.)

frame-shaped electrode was uniform above a gold island while the AFM tip effectively maintained a constant distance of ~400 nm from the gold island. An SECM image (Figure 1.35b top) was simulated with the boundary element method by using the topography image obtained from the AFM scan (Figure 1.35b, bottom). The SECM image was calculated for a small region of the substrate (white square in Figure 1.35a) to quantitatively reproduce topography and reactivity changes.

1.5.3 SCANNING ION-CONDUCTANCE MICROSCOPY

Recently, SECM was combined with SICM to enable simultaneous reactivity and topography imaging. An SECM–SICM tip was based on the nanoring electrode formed around the tip of a nanopipette (Figures 1.23c [144] and 1.31c [183]). The z position of the nanoring/nanopipette tip was adjusted during imaging to maintain constant DC ionic current through the nanopipette in the SICM mode (Figure 1.36a) [183]. The ionic current was determined by the tip–substrate distance independent of substrate reactivity. Thus, the tip followed the topography of the substrate surface to give a constant-distance topography image. Figure 1.36b shows the SICM topography image of 180-nm-wide trenches with a spacing of 875 nm as milled by FIB technology into a gold film on a glass substrate. The topography image confirmed that the trenches were lower than the original Au surface. By contrast, the trenches in the image were significantly broader than their actual width, although actual trenches were narrower than the diameter of the nanopipette opening (~100 nm).

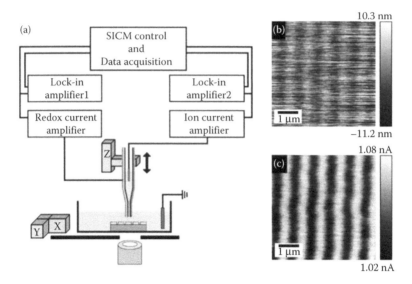

FIGURE 1.36 (a) SECM–SICM imaging instrumentation. (b) SICM and (c) SECM images of the 180-nm-wide trenches milled into a gold film on a glass substrate in 10 mM $Ru(NH_3)_6^{3+}$ and 100 mM KNO_3. (Reprinted with permission from Comstock, D. J. et al., Integrated ultra-microelectrode-nanopipet probe for concurrent scanning electrochemical microscopy and scanning ion conductance microscopy, *Anal. Chem.*, Vol. 82, 2010: pp. 1270–1276. Copyright 2010, American Chemical Society.)

This result suggests that the spatial resolution was compromised by a large tip–substrate distance. In fact, the trenches looked similarly broad in the SECM image as obtained by monitoring $Ru(NH_3)_6^{3+}$ reduction at the nanoring electrode with an even larger effective diameter of ~600 nm (Figure 1.36c). Importantly, the contrast of this constant-distance SECM image mainly reflected substrate reactivity. The tip current was lower over the less reactive trenches than over the surrounding Au surface. Noticeably, a comparison of the SICM image with the SECM image shows a lateral offset of ~200 nm, which is consistent with a center-to-center spacing between the nanopipette opening and the integrated nanoring electrode (Figure 1.31c).

The combined SECM–SICM technique was also applied to obtain the nanoscale electrochemical and topography images of various substrates including biological cells [141,184]. The SICM and SECM images of nanopores with a pore radius of ~100 nm in a polyethylene terephthalate membrane are shown in Figure 1.37a,b, respectively [141]. The images were obtained by employing an asymmetric double-barrel nanopipette with one barrel filled with pyrolytic carbon (Figure 1.22d). The effective radius of the carbon nanoelectrode was estimated to be 28 nm from SECM approach curves. The feedback current based on FcMeOH oxidation at the carbon tip was monitored to obtain the electrochemical image. Simultaneously, the open channel was used for ion-current monitoring in SICM topography imaging to maintain a constant ionic current, which corresponds to a distance of 30 nm from the membrane surface. When the tip was scanned over a nanopore, the tip moved toward

FIGURE 1.37 Simultaneous (a and c) SICM and (b and d) SECM images of (top) a nanoporous polyethylene terephthalate membrane and (bottom) living sensory neurons. The electrochemical images were based on FcMeOH oxidation at the carbon nanotip. (Takahashi, Y. et al., Multifunctional nanoprobes for nanoscale chemical imaging and localized chemical delivery at surfaces and interfaces, *Angew. Chem., Int. Ed.*, 2011. Vol. 50. pp. 9638–9642. Copyright 2011, Wiley-VCH Verlag GmbH & Co. KGaA. Reproduced with permission.)

the inside of a nanopore as seen in the SICM image to maintain the constant ionic current response. At the same time, redox current at the SECM tip was somehow lowered by a negative feedback effect from the pore wall although the nanopore was larger than the tip to serve as the source of FcMeOH. This behavior contrasts to a higher current response over a nanopore as observed by the constant-height imaging of silicon nanopores using a nanopipette-supported ITIES tip (Figure 1.8c). A double-barrel tip with an effective electrode radius of 240 nm was also employed for the simultaneous SECM–SICM imaging of living sensory neurons. The tall cell bodies and dendritic structures were clearly and consistently observed in both SICM and SECM images (Figure 1.37c,d, respectively). Importantly, the SECM image measures the flux of $FcCH_2OH$ across the cell membrane, whereas the SICM image represents its topography. In fact, a redox tip current of 38 pA over the cells was significantly higher than a negative feedback tip current of 23 pA over the bare petri dish when a constant tip–substrate distance was maintained in the constant-current SICM mode.

1.5.4 SPECTROELECTROCHEMISTRY

The SECM is an attractive platform for combination with spectroelectrochemical measurement. Optical-fiber-based SECM tips have been used for the screening of photocatalysts, although tip size in these studies is limited to micrometers [196]. By contrast, ECL was generated at an SECM nanotip as a nanoscale light source for near-field optical imaging [197]. Specifically, a Pt nanotip with an effective diameter of down to 155 nm was used as the stable source of ECL emission in a thin (~500 μm) layer of an aqueous solution of 15 mM Ru(bipyridine)$_3^{2+}$ and 100 mM tri-n-propylamine (Figure 1.38a). The thin water layer facilitates the control of the tip–substrate separation within a near field regime by measuring shear force from the tip attached to a quartz tuning fork. A nanotip was prepared by electrochemically

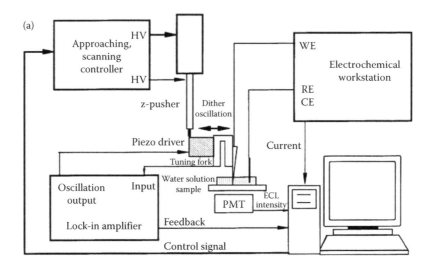

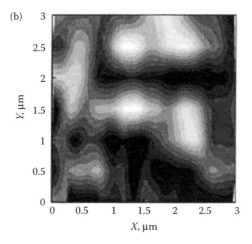

FIGURE 1.38 (a) Block diagram of the apparatus used in the ECL imaging experiments. (b) ECL near-field image of a test sample containing submicrometer holes. (Reprinted with permission from Zu, Y. et al., Scanning optical microscopy with an electrogenerated chemiluminescent light source at a nanometer tip, *Anal. Chem.*, Vol. 73, 2001: pp. 2153–2156. Copyright 2001, American Chemical Society.)

etching a Pt wire, which was further coated with an anodic electrophoretic paint and heated at ~160°C to expose a conical-shaped tip. Figure 1.38b is an ECL image of a test substrate as obtained using a 172-nm-diameter (effective) tip. The test substrate was made by employing polystyrene latex spheres with a diameter of ~0.48 μm as a template, which was randomly distributed on a glass slide and then coated with a 100-nm-thick layer of aluminum by vacuum evaporation. The beads were removed by sonication in dichloromethane to form submicrometer-sized transparent holes. In Figure 1.38b, these holes were imaged optically with submicrometer resolution by detecting the ECL photons that were emitted from the nanotip and transmitted through the holes to the underlying PMT detector. The ECL image clearly shows that some of the submicrometer-sized beads were originally isolated from one another while others clumped together. A near-field optical resolution was achieved in this study; the lateral spatial resolution was approximately one third of the emission wavelength maximum of 645 nm for the Ru(bipyridine)$_3^{2+}$/tri-n-propylamine system. Further improvement of spatial resolution with a smaller tip requires a more sensitive and low-background photon detector. Highly sensitive photon detection will be possible with an SECM setup as shown by the combination of SECM with single-molecule spectroscopy [198].

1.6 PERSPECTIVES

The transformation of SECM from micrometer scale to nanometer scale presents great opportunities to enable a wide range of its applications in nanoscience and nanotechnology. The nanoscale SECM imaging of various single nano-entities such as nanopores, nanotubes, and nanoparticles is envisioned as exciting future applications with high and broad significance in biology and materials science. These nano-entities have been studied as ensembles or individually with lower resolution by using microscale SECM [9]. In comparison, nanoscale SECM will provide higher-resolution information about the reactivity of the individual nano-entities. In addition, higher mass transport conditions of nanoscale SECM will provide kinetic information about heterogeneous and homogeneous reactions, which may be diffusion-limited when microscale SECM or other electrochemical techniques are employed. Because of this intrinsically kinetic nature, the deconvolution of reactivity and topography information is more important in nanoscale SECM. Fortunately, a few promising approaches are possible or realistic for nanoscale SECM in contrast to the microscale counterpart. Advantageously, the fast steady-state and quasi-steady-state responses of nanoscale SECM will significantly shorten an imaging time, thereby enabling us to obtain reactivity and topography images within a much shorter time scale even as a movie. Eventually, the quantitative analysis of SECM images is required to determine substrate reactivity as a rate constant. An instrument for nanoscale SECM with the ability to simultaneously obtain quantitative reactivity and topography images will be highly attractive for its users. The commercialization of such an instrument will be the important milestone that represents the reliability, maturity, and success of this powerful and interesting technique.

The establishment of nanoscale SECM as a reliable and quantitative electrochemical method requires further technological development and fundamental understanding of nanoscale processes at interfaces. It was recognized only recently that a metal nanotip can be easily damaged at the nanometer scale (Section 1.4.3) [124]. The nanoscale tip damage was suggested by quantitatively analyzing SECM data by using solid theories and was eventually confirmed by the visualization of a damaged tip. Unexpectedly and resentfully, this lesson exemplifies the importance of the quantitative and visual characterization of an individual "nanostructure." This finding also urges electrochemists to reappraise the electrochemical measurements of nanosystems reported or attempted in the past two decades including not only nanoelectrodes but also the nanometer-thick films attached to micrometer-sized electrodes. In addition to this rather physical problem, some chemical problems must be solved for nanoscale electrochemical measurements. Readers might have noticed that only certain redox mediators, for example, $Ru(NH_3)_6^{3+}$ and ferrocene derivatives, have been used for nanoscale SECM experiments. In fact, some redox mediators do not behave well at metal nanotips [83,92]. Moreover, metal [83,92] and carbon [29] nanotips cannot approach a flat macroscopic substrate as close as expected from Equation 1.4, especially when their diameters become less than 100 nm. Origins of these limitations may be related to nanoelectrode contamination [161] but are not well understood, which represents the current inadequacy of our fundamental understanding of interfacial (electro)chemical processes at the nanoscale. Importantly, nanoscale processes are ubiquitous and paramount as the individual steps of the overall process at the interfaces with any size.

ACKNOWLEDGMENTS

This work was supported by the National Institutes of Health (GM073439) and the National Science Foundation (CHE-1213452).

REFERENCES

1. Liu, H.-Y., F.-R. F. Fan, C. W. Lin, and A. J. Bard, Scanning electrochemical and tunneling ultramicroelectrode microscope for high-resolution examination of electrode surfaces in solution, *J. Am. Chem. Soc.*, Vol. 108, 1986: pp. 3838–3839.
2. Bard, A. J., F.-R. F. Fan, J. Kwak, and O. Lev, Scanning electrochemical microscopy. Introduction and principles, *Anal. Chem.*, Vol. 61, 1989: pp. 132–138.
3. Kwak, J., and A. J. Bard, Scanning electrochemical microscopy. Apparatus and two-dimensional scans of conductive and insulating substrates, *Anal. Chem.*, Vol. 61, 1989: pp. 1794–1799.
4. Lee, C., C. J. Miller, and A. J. Bard, Scanning electrochemical microscopy: Preparation of submicrometer electrodes, *Anal. Chem.*, Vol. 63, 1991: pp. 78–83.
5. Mirkin, M. V., F.-R. F. Fan, and A. J. Bard, Direct electrochemical measurements inside a 2000 angstrom thick polymer film by scanning electrochemical microscopy, *Science*, Vol. 257, 1992: pp. 364–366.
6. Mirkin, M. V., F.-R. F. Fan, and A. J. Bard, Scanning electrochemical microscopy. Part 13. Evaluation of the tip shapes of nanometer size microelectrodes, *J. Electroanal. Chem.*, Vol. 328, 1992: pp. 47–62.
7. Bard, A. J., and M. V. Mirkin, Eds., *Scanning Electrochemical Microscopy*, Marcel Dekker, New York, 2001.

8. *Visualizing Chemistry: The Progress and Promise of Advanced Chemical Imaging*, National Academies Press, Washington, DC, 2006.

9. Amemiya, S., Scanning electrochemical microscopy of nanopores, nanocarbons, and nanoparticles, in *Nanoelectrochemistry*, Mirkin M. V. and S. Amemiya, Eds., Taylor & Francis, Boca Raton, FL, 2015: pp. 621–653.

10. Amemiya, S., A. J. Bard, F.-R. F. Fan, M. V. Mirkin, and P. R. Unwin, Scanning electrochemical microscopy, *Annu. Rev. Anal. Chem.*, Vol. 1, 2008: pp. 95–131.

11. Mirkin, M. V., W. Nogala, J. Velmurugan, and Y. Wang, Scanning electrochemical microscopy in the 21st century. Update 1: Five years after, *Phys. Chem. Chem. Phys.*, Vol. 13, 2011: pp. 21196–21212.

12. Bard, A. J., and M. V. Mirkin, Eds., *Scanning Electrochemical Microscopy*. 2nd ed., CRC Press, Boca Raton, FL.

13. http://www.nano.gov/nanotech-101/what/definition.

14. Kwak, J., and A. J. Bard, Scanning electrochemical microscopy. Theory of the feedback mode, *Anal. Chem.*, Vol. 61, 1989: pp. 1221–1227.

15. Slevin, C. J., N. J. Gray, J. V. Macpherson, M. A. Webb, and P. R. Unwin, Fabrication and characterization of nanometer-sized platinum electrodes for voltammetric analysis and imaging, *Electrochem. Commun.*, Vol. 1, 1999: pp. 282–288.

16. Baltes, N., L. Thouin, C. Amatore, and J. Heinze, Imaging concentration profiles of redox-active species with nanometric amperometric probes: Effect of natural convection on transport at microdisk electrodes, *Angew. Chem., Int. Ed.*, Vol. 43, 2004: pp. 1431–1435.

17. Nioradze, N., J. Kim, and S. Amemiya, Quasi-steady-state voltammetry of rapid electron transfer reactions at the macroscopic substrate of the scanning electrochemical microscope, *Anal. Chem.*, Vol. 83, 2011: pp. 828–835.

18. Bard, A. J., M. V. Mirkin, P. R. Unwin, and D. O. Wipf, Scanning electrochemical microscopy. 12. Theory and experiment of the feedback mode with finite heterogeneous electron-transfer kinetics and arbitrary substrate size, *J. Phys. Chem.*, Vol. 96, 1992: pp. 1861–1868.

19. Wipf, D. O., and A. J. Bard, Scanning electrochemical microscopy. VII. Effect of heterogeneous electron-transfer rate at the substrate on the tip feedback current, *J. Electrochem. Soc.*, Vol. 138, 1991: pp. 469–474.

20. Xiong, H., J. Guo, and S. Amemiya, Probing heterogeneous electron transfer at an unbiased conductor by scanning electrochemical microscopy in the feedback mode, *Anal. Chem.*, Vol. 79, 2007: pp. 2735–2744.

21. Xiong, H., D. A. Gross, J. Guo, and S. Amemiya, Local feedback mode of scanning electrochemical microscopy for electrochemical characterization of one-dimensional nanostructure: Theory and experiment with nanoband electrode as model substrate, *Anal. Chem.*, Vol. 78, 2006: pp. 1946–1957.

22. Xiong, H., J. Kim, E. Kim, and S. Amemiya, Scanning electrochemical microscopy of one-dimensional nanostructure: Effects of nanostructure dimensions on the tip feedback current under unbiased conditions, *J. Electroanal. Chem.*, Vol. 629, 2009: pp. 78–86.

23. Kim, E., J. Kim, and S. Amemiya, Spatially resolved detection of a nanometer-scale gap by scanning electrochemical microscopy, *Anal. Chem.*, Vol. 81, 2009: pp. 4788–4791.

24. Kim, J., H. Xiong, M. Hofmann, J. Kong, and S. Amemiya, Scanning electrochemical microscopy of individual single-walled carbon nanotubes, *Anal. Chem.*, Vol. 82, 2010: pp. 1605–1607.

25. Laforge, F. O., J. Velmurugan, Y. Wang, and M. V. Mirkin, Nanoscale imaging of surface topography and reactivity with the scanning electrochemical microscope, *Anal. Chem.*, Vol. 81, 2009: pp. 3143–3150.

26. Wei, C., and A. J. Bard, Scanning electrochemical microscopy XXIX. *In situ* monitoring of thickness changes of thin films on electrodes, *J. Electrochem. Soc.*, Vol. 142, 1995: pp. 2523–2527.

27. Wipf, D. O., and A. J. Bard, Scanning electrochemical microscopy. 15. Improvements in imaging via tip-position modulation and lock-in detection, *Anal. Chem.*, Vol. 64, 1992: pp. 1362–1367.

28. Wang, Y., K. Kececi, J. Velmurugan, and M. V. Mirkin, Electron transfer/ion transfer mode of scanning electrochemical microscopy (SECM): A new tool for imaging and kinetic studies, *Chem. Sci.*, Vol. 4, 2013: pp. 3606–3616.

29. Takahashi, Y., A. I. Shevchuk, P. Novak, B. Babakinejad, J. Macpherson, P. R. Unwin, H. Shiku et al., Topographical and electrochemical nanoscale imaging of living cells using voltage-switching mode scanning electrochemical microscopy, *Proc. Natl. Acad. Sci. USA*, Vol. 109, 2012: pp. 11540–11545.

30. Lee, Y., Z. Ding, and A. J. Bard, Combined scanning electrochemical/optical microscopy with shear force and current feedback, *Anal. Chem.*, Vol. 74, 2002: pp. 3634–3643.

31. Bard, A. J., G. H. Denuault, R. A. Friesner, B. C. Dornblaser, and L. S. Tuckerman, Scanning electrochemical microscopy: Theory and application of the transient (chronoamperometric) SECM response, *Anal. Chem.*, Vol. 63, 1991: pp. 1282–1288.

32. Lazenby, R., K. McKelvey, M. Peruffo, M. Baghdadi, and P. Unwin, Nanoscale intermittent contact-scanning electrochemical microscopy, *J. Solid State Electrochem.*, Vol. 17, 2013: pp. 2979–2987.

33. Katemann, B. B., A. Schulte, and W. Schuhmann, Constant-distance mode scanning electrochemical microscopy. Part II: High-resolution SECM imaging employing Pt nanoelectrodes as miniaturized scanning probes, *Electroanalysis*, Vol. 16, 2004: pp. 60–65.

34. Yamada, H., Y. Ikuta, T. Koike, and T. Matsue, Fabrication of a shear force-based ion-selective capillary probe for scanning electrochemical microscopy, *Chem. Lett.*, Vol. 37, 2008: pp. 392–393.

35. Tefashe, U. M., and G. Wittstock, Quantitative characterization of shear force regulation for scanning electrochemical microscopy, *C. R. Chim.*, Vol. 16, 2013: pp. 7–14.

36. Fan, F.-R. F., and A. J. Bard, Imaging biological macromolecules on mica in humid air by scanning electrochemical microscopy, *Proc. Natl. Acad. Sci. USA*, Vol. 96, 1999: pp. 14222–14227.

37. Guckenberger, R., M. Heim, G. Cevc, H. Knapp, W. Wiegräbe, and A. Hillebrand, Scanning tunneling microscopy of insulators and biological specimens based on lateral conductivity of ultrathin water films, *Science*, Vol. 266, 1994: pp. 1538–1540.

38. Fan, F.-R. F., and A. J. Bard, STM on wet insulators: Electrochemistry or tunneling? *Science*, Vol. 270, 1995: pp. 1849–1852.

39. Forouzan, F., and A. J. Bard, Evidence for faradaic processes in scanning probe microscopy on mica in humid air, *J. Phys. Chem. B*, Vol. 101, 1997: pp. 10876–10879.

40. Shevchuk, A. I., G. I. Frolenkov, D. Sánchez, P. S. James, N. Freedman, M. J. Lab, R. Jones, D. Klenerman, and Y. E. Korchev, Imaging proteins in membranes of living cells by high-resolution scanning ion conductance microscopy, *Angew. Chem., Int. Ed.*, Vol. 45, 2006: pp. 2212–2216.

41. Korchev, Y. E., Y. A. Negulyaev, C. R. W. Edwards, I. Vodyanoy, and M. J. Lab, Functional localization of single active ion channels on the surface of a living cell, *Nat. Cell. Biol.*, Vol. 2, 2000: pp. 616–619.

42. Shen, M., R. Ishimatsu, J. Kim, and S. Amemiya, Quantitative imaging of ion transport through single nanopores by high-resolution scanning electrochemical microscopy, *J. Am. Chem. Soc.*, Vol. 134, 2012: pp. 9856–9859.

43. Striemer, C. C., T. R. Gaborski, J. L. McGrath, and P. M. Fauchet, Charge- and size-based separation of macromolecules using ultrathin silicon membranes, *Nature*, Vol. 445, 2007: pp. 749–753.

44. Snyder, J. L., J. Getpreecharsawas, D. Z. Fang, T. R. Gaborski, C. C. Striemer, P. M. Fauchet, D. A. Borkholder, and J. L. McGrath, High-performance, low-voltage electroosmotic pumps with molecularly thin silicon nanomembranes, *Proc. Natl. Acad. Sci. USA*, Vol. 110, 2013: pp. 18425–18430.

45. Bath, B. D., H. S. White, and E. R. Scott, Imaging molecular transport across membranes, in *Scanning Electrochemical Microscopy*, Bard A. J. and M. V. Mirkin, Eds. 2001, Marcel Dekker, New York, pp. 343–395.

46. Uitto, O. D., and H. S. White, Scanning electrochemical microscopy of membrane transport in the reverse imaging mode, *Anal. Chem.*, Vol. 73, 2001: pp. 533–539.

47. Uitto, O. D., H. S. White, and K. Aoki, Diffusive-convective transport into a porous membrane. A comparison of theory and experiment using scanning electrochemical microscopy operated in reverse imaging mode, *Anal. Chem.*, Vol. 74, 2002: pp. 4577–4582.

48. Lee, S., Y. Zhang, H. S. White, C. C. Harrell, and C. R. Martin, Electrophoretic capture and detection of nanoparticles at the opening of a membrane pore using scanning electrochemical microscopy, *Anal. Chem.*, Vol. 76, 2004: pp. 6108–6115.

49. Ervin, E. N., H. S. White, and L. A. Baker, Alternating current impedance imaging of membrane pores using scanning electrochemical microscopy, *Anal. Chem.*, Vol. 77, 2005: pp. 5564–5569.

50. Ervin, E. N., H. S. White, L. A. Baker, and C. R. Martin, Alternating current impedance imaging of high-resistance membrane pores using a scanning electrochemical microscope. Application of membrane electrical shunts to increase measurement sensitivity and image contrast, *Anal. Chem.*, Vol. 78, 2006: pp. 6535–6541.

51. White, R. J., and H. S. White, Influence of electrophoresis waveforms in determining stochastic nanoparticle capture rates and detection sensitivity, *Anal. Chem.*, Vol. 79, 2007: pp. 6334–6340.

52. McKelvey, K., M. E. Snowden, M. Peruffo, and P. R. Unwin, Quantitative visualization of molecular transport through porous membranes: Enhanced resolution and contrast using intermittent contact-scanning electrochemical microscopy, *Anal. Chem.*, Vol. 83, 2011: pp. 6447–6454.

53. Proksch, R., R. Lal, P. K. Hansma, D. Morse, and G. Stucky, Imaging the internal and external pore structure of membranes in fluid: Tappingmode scanning ion conductance microscopy, *Biophys. J.*, Vol. 71, 1996: pp. 2155–2157.

54. Böcker, M., S. Muschter, E. K. Schmitt, C. Steinem, and T. E. Schäffer, Imaging and patterning of pore-suspending membranes with scanning ion conductance microscopy, *Langmuir*, Vol. 25, 2009: pp. 3022–3028.

55. Chen, C.-C., M. A. Derylo, and L. A. Baker, Measurement of ion currents through porous membranes with scanning ion conductance microscopy, *Anal. Chem.*, Vol. 81, 2009: pp. 4742–4751.

56. Chen, C.-C., and L. A. Baker, Effects of pipette modulation and imaging distances on ion currents measured with scanning ion conductance microscopy (SICM), *Analyst*, Vol. 136, 2011: pp. 90–97.

57. Chen, C.-C., Y. Zhou, and L. A. Baker, Single-nanopore investigations with ion conductance microscopy, *ACS Nano*, Vol. 5, 2011: pp. 8404–8411.

58. Zhou, Y., C.-C. Chen, and L. A. Baker, Heterogeneity of multiple-pore membranes investigated with ion conductance microscopy, *Anal. Chem.*, Vol. 84, 2012: pp. 3003–3009.

59. Takahashi, Y., A. I. Shevchuk, P. Novak, Y. Zhang, N. Ebejer, J. V. Macpherson, P. R. Unwin et al., Multifunctional nanoprobes for nanoscale chemical imaging and localized chemical delivery at surfaces and interfaces, *Angew. Chem., Int. Ed.*, Vol. 50, 2011: pp. 9638–9642.

60. Morris, C. A., C.-C. Chen, and L. A. Baker, Transport of redox probes through single pores measured by scanning electrochemical-scanning ion conductance microscopy (SECM–SICM), *Analyst*, Vol. 2012: pp. 2933–2938.

61. Macpherson, J. V., and P. R. Unwin, Combined scanning electrochemical-atomic force microscopy, *Anal. Chem.*, Vol. 72, 2000: pp. 276–285.

62. Macpherson, J. V., C. E. Jones, A. L. Barker, and P. R. Unwin, Electrochemical imaging of diffusion through single nanoscale pores, *Anal. Chem.*, Vol. 74, 2002: pp. 1841–1848.

63. Gardner, C. E., P. R. Unwin, and J. V. Macpherson, Correlation of membrane structure and transport activity using combined scanning electrochemical–atomic force microscopy, *Electrochem. Commun.*, Vol. 7, 2005: pp. 612–618.

64. Kim, J., A. Izadyar, M. Shen, R. Ishimatsu, and S. Amemiya, Ion permeability and ion-induced permeabilization of the nuclear pore complex: Scanning electrochemical and fluorescence microsocpy studies, *Anal. Chem.*, Vol. 86, 2014: pp. 2090–2098.

65. Sun, P., F. O. Laforge, T. P. Abeyweera, S. A. Rotenberg, J. Carpino, and M. V. Mirkin, Nanoelectrochemistry of mammalian cells, *Proc. Natl. Acad. Sci. USA*, Vol. 105, 2008: pp. 443–448.

66. Bergner, S., J. Wegener, and F. M. Matysik, Monitoring passive transport of redox mediators across a confluent cell monolayer with single-cell resolution by means of scanning electrochemical microscopy, *Anal. Methods*, Vol. 4, 2012: pp. 623–629.

67. Bergner, S., J. Wegener, and F.-M. Matysik, Simultaneous imaging and chemical attack of a single living cell within a confluent cell monolayer by means of scanning electrochemical microscopy, *Anal. Chem.*, Vol. 83, 2011: pp. 169–174.

68. Wang, Y. X., J. M. Noel, J. Velmurugan, W. Nogala, M. V. Mirkin, C. Lu, M. G. Collignon, F. Lemaitre, and C. Amatore, Nanoelectrodes for determination of reactive oxygen and nitrogen species inside murine macrophages, *Proc. Natl. Acad. Sci. USA*, Vol. 109, 2012: pp. 11534–11539.

69. Hu, K., Y. Gao, Y. Wang, Y. Yu, X. Zhao, S. Rotenberg, E. Gökmeşe, M. Mirkin, G. Friedman, and Y. Gogotsi, Platinized carbon nanoelectrodes as potentiometric and amperometric SECM probes, *J. Solid State Electrochem.*, Vol. 17, 2013: pp. 2971–2977.

70. Sun, T., Y. Yu, B. J. Zacher, and M. V. Mirkin, Scanning electrochemical microscopy of individual catalytic nanoparticles, *Angew. Chem., Int. Ed.*, Vol. 53, 2014: pp. 14120–14123.

71. Bard, A. J., and F.-R. F. Fan, Electrochemical detection of single molecules, *Acc. Chem. Res.*, Vol. 29, 1996: pp. 572–578.

72. Rassaei, L., P. S. Singh, and S. G. Lemay, Lithography-based nanoelectrochemistry, *Anal. Chem.*, Vol. 83, 2011: pp. 3974–3980.

73. Lemay, S. G., S. Kang, K. Mathwig, and P. S. Singh, Single-molecule electrochemistry: Present status and outlook, *Acc. Chem. Res.*, Vol. 46, 2012: pp. 369–377.

74. Zevenbergen, M. A. G., B. L. Wolfrum, E. D. Goluch, P. S. Singh, and S. G. Lemay, Fast electron-transfer kinetics probed in nanofluidic channels, *J. Am. Chem. Soc.*, Vol. 131, 2009: pp. 11471–11477.

75. Wei, C., and A. J. Bard, Scanning electrochemical microscopy. 31. *In-situ* monitoring of thickness changes of thin-films on electrodes, *J. Electrochem. Soc.*, Vol. 142, 1995: pp. 2523–2527.

76. Hochstetler, S. E., M. Puopolo, S. Gustincich, E. Raviola, and R. M. Wightman, Real-time amperometric measurements of zeptomole quantities of dopamine released from neurons, *Anal. Chem.*, Vol. 72, 2000: pp. 489–496.

77. Watkins, J. J., J. Chen, H. S. White, H. D. Abruna, E. Maisonhaute, and C. Amatore, Zeptomole voltammetric detection and electron-transfer rate measurements using platinum electrodes of nanometer dimensions, *Anal. Chem.*, Vol. 75, 2003: pp. 3962–3971.

78. Fan, F.-R. F., and A. J. Bard, Electrochemical detection of single molecules, *Science*, Vol. 267, 1995: pp. 871–874.

79. Fan, F.-R. F., J. Kwak, and A. J. Bard, Single molecule electrochemistry, *J. Am. Chem. Soc.*, Vol. 118, 1996: pp. 9669–9675.

80. Sun, P., and M. V. Mirkin, Electrochemistry of individual molecules in zeptoliter volumes, *J. Am. Chem. Soc.*, Vol. 130, 2008: pp. 8241–8250.

81. Sun, P., and M. V. Mirkin, Scanning electrochemical microscopy with slightly recessed nanotips, *Anal. Chem.*, Vol. 79, 2007: pp. 5809–5816.

82. Mirkin, M. V., and A. J. Bard, Simple analysis of quasi-reversible steady-state voltammograms, *Anal. Chem.*, Vol. 64, 1992: pp. 2293–2302.

83. Sun, P., and M. V. Mirkin, Kinetics of electron-transfer reactions at nanoelectrodes, *Anal. Chem.*, Vol. 78, 2006: pp. 6526–6534.

84. Amatore, C., and E. Maisonhaute, When voltammetry reaches nanoseconds? *Anal. Chem.*, Vol. 77, 2005: pp. 303A–311A.

85. Cornut, R., A. Bhasin, S. Lhenry, M. Etienne, and C. Lefrou, Accurate and simplified consideration of the probe geometrical defaults in scanning electrochemical microscopy: Theoretical and experimental investigations, *Anal. Chem.*, Vol. 83, 2011: pp. 9669–9975.

86. Mirkin, M. V., L. O. S. Bulhões, and A. J. Bard, Determination of the kinetic parameters for the electroreduction of C_{60} by scanning electrochemical microscopy and fast scan cyclic voltammetry, *J. Am. Chem. Soc.*, Vol. 115, 1993: pp. 201–204.

87. Mirkin, M. V., T. C. Richards, and A. J. Bard, Scanning electrochemical microscopy. 20. Steady-state measurements of the fast heterogeneous kinetics in the ferrocene/acetonitrile system, *J. Phys. Chem.*, Vol. 87, 1993: pp. 7672–7677.

88. Shen, M., N. Arroyo-Currás, and A. J. Bard, Achieving nanometer scale tip-to-substrate gaps with micrometer-size ultramicroelectrodes in scanning electrochemical microscopy, *Anal. Chem.*, Vol. 83, 2011: pp. 9082–9085.

89. Satpati, A. K., and A. J. Bard, Preparation and characterization of carbon powder paste ultramicroelectrodes as tips for scanning electrochemical microscopy applications, *Anal. Chem.*, Vol. 84, 2012: pp. 9498–9504.

90. Chang, J., K. C. Leonard, S. K. Cho, and A. J. Bard, Examining ultramicroelectrodes for scanning electrochemical microscopy by white light vertical scanning interferometry and filling recessed tips by electrodeposition of gold, *Anal. Chem.*, Vol. 84, 2012: pp. 5159–5163.

91. Bonazza, H. L., and J. L. Fernández, An efficient method for fabrication of disk-shaped scanning electrochemical microscopy probes with small glass-sheath thicknesses, *J. Electroanal. Chem.*, Vol. 650, 2010: pp. 75–81.

92. Velmurugan, J., P. Sun, and M. V. Mirkin, Scanning electrochemical microscopy with gold nanotips: The effect of electrode material on electron transfer rates, *J. Phys. Chem. C.*, Vol. 113, 2008: pp. 459–464.

93. Feldberg, S. W., and N. Sutin, Distance dependence of heterogeneous electron transfer through the nonadiabatic and adiabatic regimes, *Chem. Phys.*, Vol. 324, 2006: pp. 216–225.

94. White, R. J., and H. S. White, Electrochemistry in nanometer-wide electrochemical cells, *Langmuir*, Vol. 24, 2008: pp. 2850–2855.

95. Kucernak, A. R., P. B. Chowdhury, C. P. Wilde, G. H. Kelsall, Y. Y. Zhu, and D. E. Williams, Scanning electrochemical microscopy of a fuel-cell electrocatalyst deposited onto highly oriented pyrolytic graphite, *Electrochim. Acta*, Vol. 45, 2000: pp. 4483–4491.

96. Treutler, T. H., and G. Wittstock, Combination of an electrochemical tunneling microscope (ECSTM) and a scanning electrochemical microscope (SECM): Application for tip-induced modification of self-assembled monolayers, *Electrochim. Acta*, Vol. 48, 2003: pp. 2923–2932.

97. Sklyar, O., T. H. Treutler, N. Vlachopoulos, and G. Wittstock, The geometry of nanometer-sized electrodes and its influence on electrolytic currents and metal deposition processes in scanning tunneling and scanning electrochemical microscopy, *Surf. Sci.*, Vol. 597, 2005: pp. 181–195.

98. Elsamadisi, P., Y. Wang, J. Velmurugan, and M. V. Mirkin, Polished nanopipets: New probes for high-resolution scanning electrochemical microscopy, *Anal. Chem.*, Vol. 83, 2011: pp. 671–673.

99. Kim, J., M. Shen, N. Nioradze, and S. Amemiya, Stabilizing nanometer scale tip-to-substrate gaps in scanning electrochemical microscopy using an isothermal chamber for thermal drift suppression, *Anal. Chem.*, Vol. 84, 2012: pp. 3489–3492.

100. Rodgers, P. J., S. Amemiya, Y. Wang, and M. V. Mirkin, Nanopipet voltammetry of common ions across the liquid–liquid interface. Theory and limitations in kinetic analysis of nanoelectrode voltammograms, *Anal. Chem.*, Vol. 82, 2010: pp. 84–90.

101. Iwashita, T., W. Schmickler, and J. W. Schultze, The influence of the metal on the kinetics of outer sphere redox reactions, *Ber. Bunsenges. Phys. Chem.*, Vol. 89, 1985: pp. 138–142.

102. Cai, C., and M. V. Mirkin, Electron transfer kinetics at polarized nanoscopic liquid/liquid interfaces, *J. Am. Chem. Soc.*, Vol. 128, 2006: pp. 171–179.

103. Shen, M., and A. J. Bard, Localized electron transfer and the effect of tunneling on the rates of $Ru(bpy)_3^{2+}$ oxidation and reduction as measured by scanning electrochemical microscopy, *J. Am. Chem. Soc.*, Vol. 133, 2011: pp. 15737–15742.

104. Blackburn, G. F., H. P. Shah, J. H. Kenten, J. Leland, R. A. Kamin, J. Link, J. Peterman et al., Electrochemiluminescence detection for development of immunoassays and DNA probe assays for clinical diagnostics, *Clin. Chem.*, Vol. 37, 1991: pp. 1534–1539.

105. Zoski, C. G., C. R. Luman, J. L. Fernandez, and A. J. Bard, Scanning electrochemical microscopy. 57. SECM tip voltammetry at different substrate potentials under quasi-steady-state and steady-state conditions, *Anal. Chem.*, Vol. 79, 2007: pp. 4957–4966.

106. Ekanayake, C. B., M. B. Wijesinghe, and C. G. Zoski, Determination of heterogeneous electron transfer and homogeneous comproportionation rate constants of tetracyano-quinodimethane using scanning electrochemical microscopy, *Anal. Chem.*, Vol. 85, 2013: pp. 4022–4029.

107. Zhou, J., Y. Zu, and A. J. Bard, Scanning electrochemical microscopy. Part 39. The proton/hydrogen mediator system and its application to the study of the electrocatalysis of hydrogen oxidation, *J. Electroanal. Chem.*, Vol. 491, 2000: pp. 22–29.

108. Amemiya, S., N. Nioradze, P. Santhosh, and M. J. Deible, Generalized theory for nanoscale voltammetric measurements of heterogeneous electron-transfer kinetics at macroscopic substrates by scanning electrochemical microscopy, *Anal. Chem.*, Vol. 83, 2011: pp. 5928–5935.

109. Feldberg, S. W., Implications of Marcus–Hush theory for steady-state heterogeneous electron transfer at an inlaid disk electrode, *Anal. Chem.*, Vol. 82, 2010: pp. 5176–5183.

110. Wang, Y., J. Velmurugan, M. V. Mirkin, P. J. Rodgers, J. Kim, and S. Amemiya, Kinetic study of rapid transfer of tetraethylammonium at the 1,2-dichloroethane/water interface by nanopipet voltammetry of common ions, *Anal. Chem.*, Vol. 82, 2010: pp. 77–83.

111. Sun, P., Z. Q. Zhang, Z. Gao, and Y. H. Shao, Probing fast facilitated ion transfer across an externally polarized liquid–liquid interface by scanning electrochemical micros-copy, *Angew. Chem., Int. Ed.*, Vol. 41, 2002: pp. 3445–3448.

112. Li, F., Y. Chen, P. Sun, M. Q. Zhang, Z. Gao, D. P. Zhan, and Y. H. Shao, Investigation of facilitated ion-transfer reactions at high driving force by scanning electrochemical microscopy, *J. Phys. Chem. B*, Vol. 108, 2004: pp. 3295–3302.

113. Yuan, Y., and Y. H. Shao, Systematic investigation of alkali metal ion transfer across the micro- and nano-water/1,2-dichloroethane interfaces facilitated by dibenzo-18-crown-6, *J. Phys. Chem. B*, Vol. 106, 2002: pp. 7809–7814.

114. Bi, S., B. Liu, F.-R. F. Fan, and A. J. Bard, Electrochemical studies of guanosine in DMF and detection of its radical cation in a scanning electrochemical microscopy nanogap experiment, *J. Am. Chem. Soc.*, Vol. 127, 2005: pp. 3690–3691.

115. Tel-Vered, R., D. A. Walsh, M. A. Mehrgardi, and A. J. Bard, Carbon nanofiber electrodes and controlled nanogaps for scanning electrochemical microscopy experiments, *Anal. Chem.*, Vol. 78, 2006: pp. 6959–6966.
116. Chang, J., and A. J. Bard, Detection of the Sn(III) intermediate and the mechanism of the Sn(IV)/Sn(II) electroreduction reaction in bromide media by cyclic voltammetry and scanning electrochemical microscopy, *J. Am. Chem. Soc.*, Vol. 136, 2013: pp. 311–320.
117. Cox, J. T., and B. Zhang, Nanoelectrodes: Recent advances and new directions, *Annu. Rev. Anal. Chem.*, Vol 5, 2012: pp. 253–272.
118. Shao, Y., M. V. Mirkin, G. Fish, S. Kokotov, D. Palanker, and A. Lewis, Nanometer-sized electrochemical sensors, *Anal. Chem.*, Vol. 69, 1997: pp. 1627–1634.
119. Manual of P-2000 laser puller from Sutter Instruments.
120. Amemiya, S., Y. Wang, and M. V. Mirkin, Nanoelectrochemistry at liquid/liquid interfaces, in *Specialist Periodical Reports in Electrochemistry*, Compton R. G. and J. D. Wadhawan, Eds., Vol. 12, RSC, Cambridge, UK, 2013: pp. 1–43.
121. Brown, K. T., and D. G. Flaming, *Advanced Micropipette Techniques for Cell Physiology*, Wiley, New York, 1986.
122. Katemann, N. B., and W. Schuhmann, Fabrication and characterization of needle-type Pt-disk nanoelectrodes, *Electroanalysis*, Vol. 14, 2002: pp. 22–28.
123. Mezour, M. A., M. Morin, and J. Mauzeroll, Fabrication and characterization of laser pulled platinum microelectrodes with controlled geometry, *Anal. Chem.*, Vol. 83, 2011: pp. 2378–2382.
124. Nioradze, N., R. Chen, J. Kim, M. Shen, P. Santhosh, and S. Amemiya, Origins of nanoscale damage to glass-sealed platinum electrodes with submicrometer and nanometer size, *Anal. Chem.*, Vol. 2013: pp. 6198–6202.
125. Danis, L., M. E. Snowden, U. M. Tefashe, C. N. Heinemann, and J. Mauzeroll, Development of nano-disc electrodes for application as shear force sensitive electrochemical probes, *Electrochim. Acta*, Vol. 136, 2014: pp. 121–129.
126. Noël, J.-M., J. Velmurugan, E. Gökmeşe, and M. Mirkin, Fabrication, characterization, and chemical etching of Ag nanoelectrodes, *J. Solid State Electrochem.*, Vol. 17, 2013: pp. 385–389.
127. Li, Y., D. Bergman, and B. Zhang, Preparation and electrochemical response of 1–3 nm Pt disk electrodes, *Anal. Chem.*, Vol. 81, 2009: pp. 5496–5502.
128. Kim, J., A. Izadyar, N. Nioradze, and S. Amemiya, Nanoscale mechanism of molecular transport through the nuclear pore complex as studied by scanning electrochemical microscopy, *J. Am. Chem. Soc.*, Vol. 135, 2013: pp. 2321–2329.
129. Volkert, C. A., and A. M. Minor, Focused ion beam microscopy and micromachining, *MRS Bull.*, Vol. 32, 2007: pp. 389–395.
130. Kim, J., B.-K. Kim, S. K. Cho, and A. J. Bard, Tunneling ultramicroelectrode: Nanoelectrodes and nanoparticle collisions, *J. Am. Chem. Soc.*, Vol. 136, 2014: pp. 8173–8176.
131. Yamada, H., D. Haraguchi, and K. Yasunaga, Fabrication and characterization of a K+-selective nanoelectrode and simultaneous imaging of topography and local K+ flux using scanning electrochemical microscopy, *Anal. Chem.*, Vol. 86, 2014: pp. 8547–8552.
132. Zhu, X. Y., Y. H. Qiao, X. Zhang, S. S. Zhang, X. H. Yin, J. Gu, Y. Chen, Z. W. Zhu, M. X. Li, and Y. H. Shao, Fabrication of metal nanoelectrodes by interfacial reactions, *Anal. Chem.*, Vol. 86, 2014: pp. 7001–7008.
133. Shao, Y., and M. V. Mirkin, Voltammetry at micropipet electrodes, *Anal. Chem.*, Vol. 70, 1998: pp. 3155–3161.
134. Deyhimi, F., and J. A. Coles, Rapid silylation of a glass-surface—Choice of reagent and effect of experimental parameters on hydrophobicity, *Helv. Chim. Acta*, Vol. 65, 1982: pp. 1752–1759.

135. Munoz, J. L., F. Deyhimi, and J. A. Coles, Silanization of glass in the making of ion-sensitive microelectrodes, *J. Neurosci. Methods*, Vol. 8, 1983: pp. 231–247.

136. Cai, C. X., Y. H. Tong, and M. V. Mirkin, Probing rapid ion transfer across a nanoscopic liquid-liquid interface, *J. Phys. Chem. B*, Vol. 108, 2004: pp. 17872–17878.

137. Ishimatsu, R., J. Kim, P. Jing, C. C. Striemer, D. Z. Fang, P. M. Fauchet, J. L. McGrath, and S. Amemiya, Ion-selective permeability of a ultrathin nanopore silicon membrane as probed by scanning electrochemical microscopy using micropipet-supported ITIES tips, *Anal. Chem.*, Vol. 82, 2010: pp. 7127–7134.

138. Amemiya, S., unpublished results.

139. Kim, Y.-T., D. M. Scarnulis, and A. G. Ewing, Carbon-ring electrodes with 1-µm tip diameter, *Anal. Chem.*, Vol. 58, 1986: pp. 1782–1786.

140. McKelvey, K., B. P. Nadappuram, P. Actis, Y. Takahashi, Y. E. Korchev, T. Matsue, C. Robinson, and P. R. Unwin, Fabrication, characterization, and functionalization of dual carbon electrodes as probes for scanning electrochemical microscopy (SECM), *Anal. Chem.*, Vol. 85, 2013: pp. 7519–7526.

141. Takahashi, Y., A. I. Shevchuk, P. Novak, Y. J. Zhang, N. Ebejer, J. V. Macpherson, P. R. Unwin et al., Multifunctional nanoprobes for nanoscale chemical imaging and localized chemical delivery at surfaces and interfaces, *Angew. Chem., Int. Ed.*, Vol. 50, 2011: pp. 9638–9642.

142. Hu, K., Y. Wang, H. Cai, M. V. Mirkin, Y. Gao, G. Friedman, and Y. Gogotsi, Open carbon nanopipettes as resistive-pulse sensors, rectification sensors, and electrochemical nanoprobes, *Anal. Chem.*, Vol. 86, 2014: pp. 8897–8901.

143. Yu, Y., J.-M. Noël, M. V. Mirkin, Y. Gao, O. Mashtalir, G. Friedman, and Y. Gogotsi, Carbon pipette-based electrochemical nanosampler, *Anal. Chem.*, Vol. 86, 2014: pp. 3365–3372.

144. Thakar, R., A. E. Weber, C. A. Morris, and L. A. Baker, Multifunctional carbon nanoelectrodes fabricated by focused ion beam milling, *Analyst*, Vol. 138, 2013: pp. 5973–5982.

145. Cornut, R., and C. Lefrou, New analytical approximations for negative feedback currents with a microdisk SECM tip, *J. Electroanal. Chem.*, Vol. 604, 2007: pp. 91–100.

146. Sun, P., and M. V. Mirkin, Scanning electrochemical microscopy with slightly recessed nanotips, *Anal. Chem.*, Vol. 79, 2007: pp. 5809–5816.

147. Xiong, H., J. Guo, K. Kurihara, and S. Amemiya, Fabrication and characterization of conical microelectrode probes templated by selectively etched optical fibers for scanning electrochemical microscopy, *Electrochem. Commun.*, Vol. 6, 2004: pp. 615–620.

148. Zoski, C. G., and M. V. Mirkin, Steady-state limiting currents at finite conical microelectrodes, *Anal. Chem.*, Vol. 74, 2002: pp. 1986–1992.

149. Brown, K. T., and D. G. Flaming, *Advanced Micropipette Techniques for Cell Physiology*, Wiley, New York, Chapter 17, 1986.

150. Lefrou, C., A unified new analytical approximation for positive feedback currents with a microdisk SECM tip, *J. Electroanal. Chem.*, Vol. 592, 2006: pp. 103–112.

151. Lee, Y., S. Amemiya, and A. J. Bard, Scanning electrochemical microscopy (SECM). 41. Theory and characterization of ring electrodes, *Anal. Chem.*, Vol. 73, 2001: pp. 2261–2267.

152. Fulian, Q., A. C. Fisher, and G. Denuault, Application of the boundary element method in electrochemistry: Scanning electrochemical microscopy, *J. Phys. Chem. B*, Vol. 103, 1999: pp. 4387–4392.

153. Zoski, C. G., B. Liu, and A. J. Bard, Scanning electrochemical microscopy: Theory and characterization of electrodes of finite conical geometry, *Anal. Chem.*, Vol. 76, 2004: pp. 3646–3654.

154. Selzer, Y., and D. Mandler, Scanning electrochemical microscopy. Theory of the feedback mode for hemispherical ultramicroelectrodes: Steady-state and transient behavior, *Anal. Chem.*, Vol. 72, 2000: pp. 2383–2390.

155. Lindsey, G., S. Abercrombie, G. Denuault, S. Daniele, and E. De Faveri, Scanning electrochemical microscopy: Approach curves for sphere-cap scanning electrochemical microscopy tips, *Anal. Chem.*, Vol. 79, 2007: pp. 2952–2956.

156. Fulian, Q., A. C. Fisher, and G. Denuault, Application of the boundary element method in electrochemistry: Scanning electrochemical microscopy, part 2, *J. Phys. Chem. B*, Vol. 103, 1999: pp. 4393–4398.

157. Velmurugan, J., and M. V. Mirkin, Fabrication of nanoelectrodes and metal clusters by electrodeposition, *ChemPhysChem.*, Vol. 11, 2010: pp. 3011–3017.

158. Nogala, W., J. Velmurugan, and M. V. Mirkin, Atomic force microscopy of electrochemical nanoelectrodes, *Anal. Chem.*, Vol. 84, 2012: pp. 5192–5197.

159. Noël, J.-M., Y. Yu, and M. V. Mirkin, Dissolution of Pt at moderately negative potentials during oxygen reduction in water and organic media, *Langmuir*, Vol. 29, 2013: pp. 1346–1350.

160. Penner, R. M., M. J. Heben, T. L. Longin, and N. S. Lewis, Fabrication and use of nanometer-sized electrodes in electrochemistry, *Science*, Vol. 250, 1990: pp. 1118–1121.

161. Scherson, D. A., and Y. V. Tolmachev, Impurity effects on oxygen reduction electrocatalysis at platinum ultramicroelectrodes: A critical assessment, *Electrochem. Solid-State Lett.*, Vol. 13, 2010: pp. F1–F2.

162. Chen, S., and A. Kucernak, Electrocatalysis under conditions of high mass transport rate: Oxygen reduction on single submicrometer-sized Pt particles supported on carbon, *J. Phys. Chem. B*, Vol. 108, 2004: pp. 3262–3276.

163. Available from: http://www.physikinstrumente.com/en/products/nanopositioning/nanopositioning_basics.php.

164. Hugelmann, P., and W. Schindler, *In-situ* voltage tunneling spectroscopy at electrochemical interfaces, *J. Phys. Chem. B*, Vol. 109, 2005: pp. 6262–6267.

165. Kranz, C., Recent advancements in nanoelectrodes and nanopipettes used in combined scanning electrochemical microscopy techniques, *Analyst*, Vol. 139, 2014: pp. 336–352.

166. Macpherson, J. V., P. R. Unwin, A. C. Hillier, and A. J. Bard, *In-situ* imaging of ionic crystal dissolution using an integrated electrochemical/AFM probe, *J. Am. Chem. Soc.*, Vol. 118, 1996: pp. 6445–6452.

167. Lee, Y., and A. J. Bard, Fabrication and characterization of probes for combined scanning electrochemical/optical microscopy experiments, *Anal. Chem.*, Vol. 74, 2002: pp. 3626–3633.

168. Takahashi, Y., Y. Hirano, T. Yasukawa, H. Shiku, H. Yamada, and T. Matsue, Topographic, electrochemical, and optical images captured using standing approach mode scanning electrochemical/optical microscopy, *Langmuir*, Vol. 22, 2006: pp. 10299–10306.

169. Ueda, A., O. Niwa, K. Maruyama, Y. Shindo, K. Oka, and K. Suzuki, Neurite imaging of living PC12 cells with scanning electrochemical/near-field optical/atomic force microscopy, *Angew. Chem., Int. Ed.*, Vol. 46, 2007: pp. 8238–8241.

170. Kranz, C., G. Friedbacher, B. Mizaikoff, A. Lugstein, J. Smoliner, and E. Bertagnolli, Integrating an ultramicroelectrode in an AFM cantilever: Combined technology for enhanced information, *Anal. Chem.*, Vol. 73, 2001: pp. 2491–2500.

171. Lugstein, A., E. Bertagnolli, C. Kranz, A. Kueng, and B. Mizaikoff, Integrating micro- and nanoelectrodes into atomic force microscopy cantilevers using focused ion beam techniques, *Appl. Phys. Lett.*, Vol. 81, 2002: pp. 349–351.

172. Moon, J. S., H. Shin, B. Mizaikoff, and C. Kranz, Bitmap-assisted focused ion beam fabrication of combined atomic force scanning electrochemical microscopy probes, *J. Korean Phys. Soc.*, Vol. 51, 2007: pp. 920–924.

173. Kueng, A., C. Kranz, B. Mizaikoff, A. Lugstein, and E. Bertagnolli, Combined scanning electrochemical atomic force microscopy for tapping mode imaging, *Appl. Phys. Lett.*, Vol. 82, 2003: pp. 1592–1594.

174. Kueng, A., C. Kranz, A. Lugstein, E. Bertagnolli, and B. Mizaikoff, AFM-tip-integrated amperometric microbiosensors: High-resolution imaging of membrane transport, *Angew. Chem., Int. Ed.*, Vol. 44, 2005: pp. 3419–3422.

175. Lee, E., M. Kim, J. Seong, H. Shin, and G. Lim, An L-shaped nanoprobe for scanning electrochemical microscopy-atomic force microscopy, *Phys. Status Solidi RRL*, Vol. 7, 2013: pp. 406–409.

176. Hirata, Y., S. Yabuki, and F. Mizutani, Application of integrated SECM ultra-micro-electrode and AFM force probe to biosensor surfaces, *Bioelectrochemistry*, Vol. 63, 2004: pp. 217–224.

177. Frederix, P., P. D. Bosshart, T. Akiyama, M. Chami, M. R. Gullo, J. J. Blackstock, K. Dooleweerdt, N. F. de Rooij, U. Staufer, and A. Engel, Conductive supports for combined AFM-SECM on biological membranes, *Nanotechnology*, Vol. 19, 2008: pp. 384004.

178. Pust, S. E., M. Salomo, E. Oesterschulze, and G. Wittstock, Influence of electrode size and geometry on electrochemical experiments with combined SECM-SFM probes, *Nanotechnology*, Vol. 21, 2010: pp. 105709.

179. Salomo, M., S. E. Pust, G. Wittstock, and E. Oesterschulze, Integrated cantilever probes for SECM/AFM characterization of surfaces, *Microelectron. Eng.*, Vol. 87, 2010: pp. 1537–1539.

180. Derylo, M. A., K. C. Morton, and L. A. Baker, Parylene insulated probes for scanning electrochemical-atomic force microscopy, *Langmuir*, Vol. 27, 2011: pp. 13925–13930.

181. Burt, D. P., N. R. Wilson, J. M. R. Weaver, P. S. Dobson, and J. V. Macpherson, Nanowire probes for high resolution combined scanning electrochemical microscopy-atomic force microscopy, *Nano Lett.*, Vol. 5, 2005: pp. 639–643.

182. Wain, A. J., D. Cox, S. Q. Zhou, and A. Turnbull, High-aspect ratio needle probes for combined scanning electrochemical microscopy-atomic force microscopy, *Electrochem. Commun.*, Vol. 13, 2011: pp. 78–81.

183. Comstock, D. J., J. W. Elam, M. J. Pellin, and M. C. Hersam, Integrated ultramicro-electrode-nanopipet probe for concurrent scanning electrochemical microscopy and scanning ion conductance microscopy, *Anal. Chem.*, Vol. 82, 2010: pp. 1270–1276.

184. Takahashi, Y., A. I. Shevchuk, P. Novak, Y. Murakami, H. Shiku, Y. E. Korchev, and T. Matsue, Simultaneous noncontact topography and electrochemical imaging by SECM/SICM featuring ion current feedback regulation, *J. Am. Chem. Soc.*, Vol. 132, 2010: pp. 10118–10126.

185. Gardner, C. E., and J. V. Macpherson, Atomic force microscopy probes go electro-chemical, *Anal. Chem.*, Vol. 74, 2002: pp. 576A–584A.

186. Eifert, A., and C. Kranz, Hyphenating atomic force microscopy, *Anal. Chem.*, Vol. 86, 2014: pp. 5190–5200.

187. Jones, C. E., J. V. Macpherson, Z. H. Barber, R. E. Somekh, and P. R. Unwin, Simultaneous topographical and amperometric imaging of surfaces in air: Towards a combined scanning force-scanning electrochemical microscope (SF-SECM), *Electrochem. Commun.*, Vol. 1, 1999: pp. 55–60.

188. Macpherson, J. V., and P. R. Unwin, Noncontact electrochemical imaging with combined scanning electrochemical atomic force microscopy, *Anal. Chem.*, Vol. 73, 2001: pp. 550–557.

189. Abbou, J., C. Demaille, M. Druet, and J. Moiroux, Fabrication of submicrometer-sized gold electrodes of controlled geometry for scanning electrochemical-atomic force microscopy, *Anal. Chem.*, Vol. 74, 2002: pp. 6355–6363.

190. Ghorbal, A., F. Grisotto, J. Charlier, S. Palacin, C. Goyer, and C. Demaille, Localized electrografting of vinylic monomers on a conducting substrate by means of an inte-grated electrochemical AFM probe, *ChemPhysChem.*, Vol. 10, 2009: pp. 1053–1057.

191. Gullo, M. R., P. Frederix, T. Akiyama, A. Engel, N. F. deRooij, and U. Staufer, Characterization of microfabricated probes for combined atomic force and high-resolution scanning electrochemical microscopy, *Anal. Chem.*, Vol. 78, 2006: pp. 5436–5442.

192. Frederix, P., M. R. Gullo, T. Akiyama, A. Tonin, N. F. de Rooij, U. Staufer, and A. Engel, Assessment of insulated conductive cantilevers for biology and electrochemistry, *Nanotechnology*, Vol. 16, 2005: pp. 997–1005.

193. Wain, A. J., A. J. Pollard, and C. Richter, High-resolution electrochemical and topographical imaging using batch-fabricated cantilever probes, *Anal. Chem.*, Vol. 86, 2014: pp. 5143–5149.

194. Güell, A. G., N. Ebejer, M. E. Snowden, J. V. Macpherson, and P. R. Unwin, Structural correlations in heterogeneous electron transfer at monolayer and multilayer graphene electrodes, *J. Am. Chem. Soc.*, Vol. 134, 2012: pp. 7258–7261.

195. Sklyar, O., A. Kueng, C. Kranz, B. Mizaikoff, A. Lugstein, E. Bertagnolli, and G. Wittstock, Numerical simulation of scanning electrochemical microscopy experiments with frame-shaped integrated atomic force microscopy-SECM probes using the boundary element method, *Anal. Chem.*, Vol. 77, 2005: pp. 764–771.

196. Bard, A., H. C. Lee, K. Leonard, H. S. Park, and S. Wang, Rapid screening methods in the discovery and investigation of new photocatalyst compositions, in *Photoelectrochemical Water Splitting: Materials, Processes and Architectures*, Lewerenz H.-J. and L. Peter, Eds., Vol. 9, RSC, Cambridge, UK, 2013: pp. 132–153.

197. Zu, Y., Z. Ding, J. Zhou, Y. Lee, and A. J. Bard, Scanning optical microscopy with an electrogenerated chemiluminescent light source at a nanometer tip, *Anal. Chem.*, Vol. 73, 2001: pp. 2153–2156.

198. Boldt, F. M., J. Heinze, M. Diez, J. Petersen, and M. Borsch, Real-time pH microscopy down to the molecular level by combined scanning electrochemical microscopy/single-molecule fluorescence spectroscopy, *Anal. Chem.*, Vol. 76, 2004: pp. 3473–3481.

2 Electrochemical Applications of Scanning Ion Conductance Microscopy

Anna Weber, Wenqing Shi, and Lane A. Baker

CONTENTS

2.1 INTRODUCTION

Since the inception of scanning tunneling microscopy (STM),[1] electrochemists have sought to take advantage of scanned probe microscopy (SPM) techniques to manipulate the spatial position of an electrode with high resolution. In addition to STM, a number of SPM methods, most notably scanning electrochemical microscopy (SECM),[2,3] have been used to investigate electrochemical processes at the micro- and nanoscale. However, inherent SPM disadvantages such as poor probe control, tedious probe fabrication, and non-physiological imaging conditions can limit the application of SPM for electrochemical characterization. In this chapter, we discuss a method that overcomes these challenges called scanning ion conductance micros- copy (SICM), a scanned probe instrument capable of ion conductance measurements at the nanoscale. Specifically, advances in probe fabrication and feedback param- eters are considered, followed by discussion of recent SICM applications for electron transfer and ion current measurements.

Scanning ion conductance microscopy was originally developed by Hansma et al. in 1989 for *in situ* examination of nonconductive surfaces.[4] The scanned probe is a hollow pipette pulled to nano- or microscale dimensions and filled with electrolyte. An electrode is inserted into the probe and a reference electrode is positioned in the bath electrolyte solution. A bias applied between the two elec- trodes forces ion current to flow from solution through the pipette tip for detection. As the probe approaches a sample in the Z direction, ion current decreases as the pathway between probe and surface is restricted. The dependence of ion current magnitude on probe–surface distance (D_{ps}) is the basis for SICM's robust feedback mechanism. Topographic and ion current images are generated by recording Z position of the probe and corresponding ion conductance as the sample is scanned. SICM has developed as a multifunctional tool for analysis of diverse samples, such as biological cells,[5,6] electrochemical cells,[7] lithium ion batteries,[8] solution interfaces,[9] and permeable membranes.[10–15] Modifications of the original two elec- trode configuration have been developed to extend SICM beyond ion current mea- surements.[11,12,16,17] Hybridization of SICM with SECM[18–21] has been applied for simultaneous ion current and faradaic current detection, which will be discussed in detail herein.

2.2 PROPERTIES OF NANOPIPETTES

Scanning probe microscopy measurements are highly dependent on probe prop- erties such as geometry and size. For SICM, pipettes with tip diameters on the order of tens to hundreds of nanometers, referred to as nanopipettes, are typically used. Probe fabrication is simple, as a filament- or laser-based puller enables high- throughput production of nanopipettes with similar tip dimensions and geometries. A capillary is clamped into the puller and heat is applied to the capillary center. After a predetermined length of heat exposure, a mechanical pull separates the cap- illary into two identical pipettes. Laser puller parameters such as laser power, scan width, trip velocity, delay time, heat-on time, and hard pull strength can be varied. Modification of these parameters ultimately affects the geometry of the pipette, such as tip length, cone angle, and tip diameter. Capillaries are commercially available with a variety of options for inner and outer diameters, number of barrels, and start- ing material. Single- or double-barrel quartz or borosilicate glass pipettes are used commonly in SICM experiments. Schematics (Figure 2.1a–c) and scanning electron micrographs (Figure 2.1d–f) of different pipette types are shown. A diagram of a

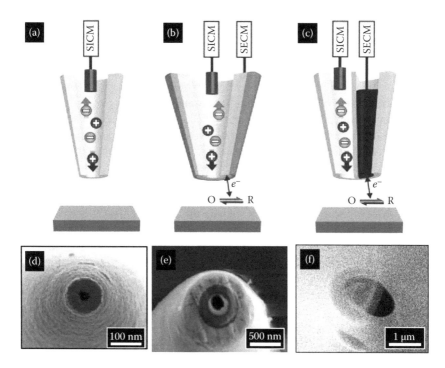

FIGURE 2.1 Cartoon illustrations of SICM probes and corresponding scanning electron micrographs. (a) Schematic of a pipette filled with electrolyte solution and pipette electrode for typical SICM experiments. (b) Schematic of a pipette coated with a conductive material (middle) and subsequently insulated with an insulating layer (outer) for simultaneous SECM-SICM measurements. (c) Schematic of a double-barrel pipette with an SICM barrel containing electrolyte and an electrode (left) and a carbon filled electrode for SECM measurement (right). (d) Scanning electron micrograph of a single barrel pipette for typical SICM experiments. (e) Scanning electron micrograph of a single-barrel, dual SECM-SICM probe. (f) Scanning electron micrograph of a double-barrel SECM-SICM probe. (Panel d reprinted from Chen, C.-C.; Zhou, Y.; Baker, L. A., *Annual Review of Analytical Chemistry* **2012**, *5*, 207–228. Panel e reprinted with permission from Takahashi, Y. et al., Simultaneous noncontact topography and electrochemical imaging by SECM/SICM featuring ion current feedback regulation. *Journal of the American Chemical Society* **2010**, *132*, 10118–10126; Copyright 2010 American Chemical Society. Panel f Takahashi, Y. et al., Multifunctional nanoprobes for nanoscale chemical imaging and localized chemical delivery at surfaces and interfaces. *Angewandte Chemie International Edition* **2011**, *50*, 9638–9642. Copyright 2011 Wiley-VCH verlas GmbH & Co. KGaA. Reproduced with permission.)

single-barrel pipette is displayed in Figure 2.1a, and an electron micrograph is shown in Figure 2.1d. Single-barrel pipettes can be modified to measure faradaic current through fabrication of an electrode around the nanopipette opening,[18–21] indicated in Figure 2.1b,e. Double-barrel pipettes, Figure 2.1c,f, has been used to increase SICM functionality; for instance, simultaneous ion current and potentiometric measurements can be made[17] or double-barrel pipettes can be used to deposit small molecules onto a surface.[22]

Nanopipettes have electrochemical properties that depend on the scale and geometry of nanopipette tips. The most notable characteristic is ion current rectification (ICR), described by Wei et al. with quartz nanopipettes.[23] Briefly, ion current response preferentially flows in one direction. A number of experimental and theoretical studies have been performed to determine the origin of the ICR effect,[24–29] and especially important treatments of the mechanism have been provided by Woermann with ICR studies through nanopores.[28,29] Under an applied trans-pore potential and at relevant pore dimensions and electrolyte concentrations, negative surface charge from deprotonated silanol groups at the nanopipette surface results in accumulation or depletion of ions, which correspondingly results in high and low conductance states. Asymmetric pore geometry generates a nonlinear current–voltage response, which results in ICR.

Nanopipettes have an overall conical geometry and thus highly localized, strong electric fields can be generated at the pipette tip, when relatively small differences in potential are applied between the pipette electrode and the reference electrode in the bath electrolyte. In comparison to the entire nanopipette body, the nanopipette tip becomes sensitive to changes in electric field or ion concentration. Ion flow in the pipette can be influenced by phenomena such as electroosmosis, electrophoresis (EP), dielectrophoresis (DEP), external pressure, or combinations of these effects.[30] Electroosmotic flow refers to a plug flow induced by the movement of the electrical double layer. If the electrolyte used in the pipette and bath solution has high conductivity, the electrical double layer thickness is often sufficiently thin such that electroosmotic flow can usually be neglected. Through careful control of applied potential, DEP trapping of proteins, nucleic acids, and small molecules has been achieved when the EP and DEP forces coexist and balance at the pipette tip.[30,31] Although DEP in a nanoscale structure (such as a nanopipette) can be used to trap, concentrate, and manipulate biomolecules, coupling nanopipettes with SICM further enables localized deposition at surfaces with high spatial resolution.[32–35] For instance, submicron patterns have been generated by controlled deposition of small quantities of reactive species, such as biotinylated and fluorophore-labeled DNA to a functionalized substrate (e.g., streptavidin) such that the delivered species are immobilized on the surface.[32]

2.3 PRINCIPLE OF OPERATION

Scanning ion conductance microscopy relies on the detection of a distance-dependent ion current for topographic and ion conductance imaging. A diagram of the typical SICM setup is shown in Figure 2.2. The probe consists of an Ag/AgCl wire back-inserted into an electrolyte-filled pipette. A second Ag/AgCl reference electrode is placed in the electrolyte bath solution. When the probe and bath come into contact and a bias is applied between the reference and pipette electrodes, ions migrate from the solution and through the pipette tip, where ion current is measured. As the probe approaches the surface, ion flow is hindered and ion current decreases. This relationship can be examined experimentally with an approach curve, where ion current is plotted as a function of D_{ps}. A distinct ion current value correlates to every D_{ps} measurement. Probe–surface distance can be precisely controlled with ion current feedback when a specific ion current value, called a set point, is indicated

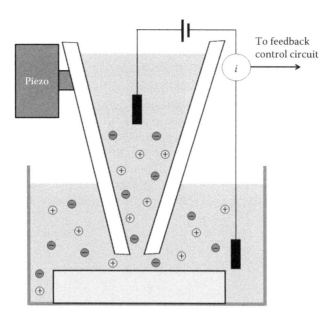

FIGURE 2.2   Schematic of typical scanning ion conductance microscopy setup. A glass or quartz nanopipette is filled with electrolyte and an Ag/AgCl wire electrode is back inserted. A reference electrode is placed in the bath electrolyte solution. To measure conductance at the pipette tip, a bias is applied between the pipette and reference electrode. Ion current detection serves as the basis for feedback control and noncontact image acquisition. As the pipette is scanned over a sample surface, a piezoelectric positioner maintains a predefined distance above the sample, in the Z direction. Vertical position changes of the pipette are recorded and topographic images are generated in this way.

to the system. A piezoelectric positioner adjusts the height of the probe whenever a deviation from the set point is detected, which enables noncontact imaging.

Ion current magnitude can be described mathematically as a function of resistance based on pipette geometry and D_{ps}.[36] Pipette resistance (R_p), shown in Equation 2.1, is dependent on pipette geometry, where h is the length of the pipette tip, κ is the conductivity of the electrolyte, r_p is the inner radius of the pipette base, and r_i is the inner radius of the pipette tip. Access resistance (R_{AC}) is the resistance measured in the separation between probe and surface and described by Equation 2.2, where r_o is outer radius of the pipette. Pipette resistance values are similar for probes of comparable geometry; therefore, the total pipette resistance is dependent on R_{AC} when pipettes of similar geometry are used. Ion current magnitude is inversely proportional to R_{AC} and R_p, as indicated in Equation 2.3.

$$R_p = \frac{h}{\kappa \pi \cdot r_p \cdot r_i} \tag{2.1}$$

$$R_{AC} \approx \frac{(3/2)\ln(r_o/r_i)}{\kappa \pi \cdot D_{ps}} \tag{2.2}$$

$$I = \frac{1}{R_{\mathrm{p}} + R_{\mathrm{AC}}} \tag{2.3}$$

Three related mathematical equations (Equations 2.4 through 2.6) have been proposed to describe the relationship between ion current magnitude and probe–surface distance.[37–39] A representation of the nanopipette tip at a surface with all relevant parameters is shown in Figure 2.3a. The parameters in each equation have been normalized, where RG represents the ratio between the outer radius (r_{o}) and the inner radius (r_{i}) of a nanopipette, RG $= (r_{\mathrm{o}}/r_{\mathrm{i}})$; r_{p} is the inner radius of the tip base; and b is the length of the tip. Subsequent equations are expressed as the normalized ion current (i/i_{∞}), as a function of the normalized probe–surface distance, $L = D_{\mathrm{ps}}/r_{\mathrm{i}}$.

$$\frac{i}{i_{\infty}} = \frac{1}{1 + ((\ln \mathrm{RG}) \cdot r_{\mathrm{p}}/2 \cdot b \cdot L)} \tag{2.4}$$

$$\frac{i}{i_{\infty}} = \frac{4b + 0.9473\pi r_0}{4b + \pi r_0(0.292 + 1.5151/L + 0.6553\exp(-2.4035)/L)} \tag{2.5}$$

$$\frac{i}{i_{\infty}} = \frac{(2.08/\mathrm{RG}^{0.358})(L - (0.145/\mathrm{RG})) + 1.585}{(2.08/\mathrm{RG}^{0.358})(L + 0.0023\mathrm{RG}) + 1.57 + (\ln \mathrm{RG}/L) + (2/\pi \mathrm{RG})\ln(1 + (\pi \mathrm{RG}/2L))} \tag{2.6}$$

In Equation 2.4, the normalized ion current is only controlled by the pipette geometry and D_{ps}.[37] In Equation 2.5, the effect of varying pipette geometry and

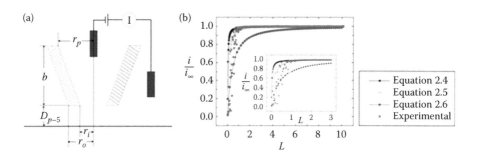

FIGURE 2.3 (a) Schematic diagram of a single barrel pipette which indicates parameters used to define the relationship between ion current magnitude and probe–surface distance. The parameters include inner radius of the tip base (r_{p}), length of the pipette tip (b), probe–surface distance (d), inner radius of the tip opening (r_{i}), and outer radius of the tip opening (r_{o}). (b) Experimentally generated approach curve (non-modulated mode) with theoretical approach curves generated from Equations 2.3 through 2.5. The horizontal axis represents the normalized probe–surface distance and the vertical axis represents normalized ion current. As the probe–surface distance increases, ion current increases until a steady state is achieved.

the subsequent influence on SICM imaging were explored. Simulated results have demonstrated that larger cone angle and RG values will make SICM more sensitive to probe–surface distance.[38] Equation 2.6 is the analytical approximation for steady-state negative feedback currents in SECM experiments for a micro disk electrode proposed by Cornut, which also includes RG.[39] Approach curves for a common pipette geometry are shown in Figure 2.3b, as well as an experimental approach curve.

2.4   FEEDBACK MODES IN SICM

2.4.1   Non-Modulated Mode (DC Feedback Mode)

Three feedback mechanisms have been developed for SICM: non-modulated (DC feedback) mode, modulated (AC feedback) mode, and iterative approach modes. The original SICM configuration was operated in DC feedback mode, shown in Figure 2.4a, where a constant potential is applied between the pipette and the reference electrodes, and the feedback mechanism is controlled by changes in ion current.[4,40] A DC approach curve (Figure 2.4d) can be divided into two separate regions where (a) the pipette tip is far from the sample surface and (b) a sharp drop in ion current is observed. In portion, (a) the pipette tip is located sufficiently far from the sample surface and the steady-state ion current is at a maximum value, independent of D_{ps}. In portion (b), R_{AC} increases as D_{ps} decreases, which results in reduced ion current magnitude. An important consideration for current sensitivity in DC mode is RG value. Typically, an RG of 1.1 maintains D_{ps} on the order of r_i.[38] A limitation of non-modulated mode is that SICM operation with high-aspect ratio features can be difficult because factors such as DC drift, partial blockage of the pipette tip, or change in ionic strength of the bath solution can result in poor feedback stability. Improved feedback modes have been introduced to address these challenges.

2.4.2   Distance-Modulated Mode (AC Feedback Mode)

To avoid instrumental artifacts in tip positioning caused by drift and other factors that complicate DC feedback mode, distance-modulated (AC feedback) mode was developed[41–46] and is shown in Figure 2.4b. In AC feedback mode, the pipette oscillates vertically at a fixed D_{ps} to generate an alternating signal. When the modulated pipette is approached to a sample surface, an oscillation in R_{AC} is produced. Combination of the AC and existing DC signals creates a modulated ion current response. An AC approach curve, shown in Figure 2.4e, is obtained by plotting the AC component of the modulated current as a function of the normalized probe–surface distance (d/r_i). Unlike non-modulated mode, the feedback signal increases as D_{ps} decreases. When probe–surface distance in AC feedback mode is maintained at a distance on the order of one pipette inner radius, a modulated current that is 0.2%–3% of the DC current signal is generated.[42] Whereas DC feedback mode is affected by changes in solution conductivity, nominal changes in ionic strength with AC feedback mode cause modulated current to only change 1%–2%.[43] In addition, better signal to noise can be attained because the system measures ion current versus

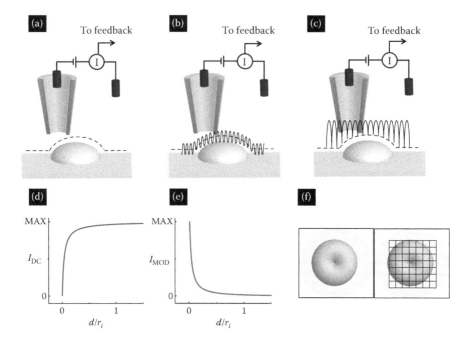

FIGURE 2.4 Schematic diagrams of SICM feedback modes and approach curves. (a) Non-modulated (DC) feedback. Application of a constant potential between reference and pipette electrodes enables the pipette to smoothly scan over the sample surface. The resulting approach curve (d) indicates the distance-dependent relationship of ion current on D_{ps}. As the pipette tip approaches the surface, ion diffusion to the tip is inhibited and ion current decreases. (b) Modulated (AC) feedback. The pipette is modulated at a fixed distance and frequency above the surface to generate an AC current, which provides a more robust feedback signal. The resulting approach curve (e) indicates the ion current-D_{ps} relationship. In this instance, ion current is at a maximum value close to the surface. (c) Repeated approach. In backstep, hopping, and standing approach modes the probe is repeatedly approached and retracted to the surface to generate a topographic image. (f) General method for repeated approach imaging. The sample is sectioned into equal segments and the pipette is approached into each unit. For low-resolution imaging, the substrate has fewer sampling points than a high-resolution image.

a zero background such that oscillations far from the surface do not influence ion conductance measurements.

2.4.3 Repetitive Approach Modes: Picking/Backstep/ Hopping/Standing Approach Modes

To image samples with complicated surface features, feedback modes which use iterative approach and retraction of the pipette have been developed[47-50] and a schematic is shown in Figure 2.4c. Generally, the pipette is approached to and retracted from a surface multiple times in a confined sample area rather than scanned. Data from approach curves taken at each imaging point are correlated to produce topographic and ion conductance images. A number of similar methods have been introduced and subtle differences among the various modes, such as set point and retraction step, differentiate the techniques. Inspiration for approach-and-retract modes in SICM can be found in picking mode for SECM reported by Heinze and coworkers.[47] To operate SECM in this mode, an ultramicroelectrode (UME) probe is approached to the sample with a velocity of 50–70 μm/s and stopped once a predefined faradaic current measurement is reached. The Z position of the probe is recorded and the probe is withdrawn, positioned over a new sample area and the process is repeated. After sufficient points on the surface have been "picked" a topographic map is generated. Logistically, the approach of the probe must be quicker than the transport rate of the redox mediator at the surface to avoid electrochemical interaction of the UME with the sample. Therefore, the change in probe current is a result of convection and the current contribution due to electrochemical surface properties is negligible. The ability of picking mode to decouple electrochemical and topographic signals was demonstrated, when the topography of a platinum sheet and PVC tape were measured independent of their conductive surface properties.[47]

The first iterative approach mode developed for SICM was backstep mode, reported by Dietzel and coworkers.[50–54] The motivation for backstep mode development was to eliminate electrode drift over time, as well as to improve the capability of SICM to image complex samples. The major difference between picking and backstep modes is that D_{ps} is regulated through application of current pulses at the pipette tip with the latter technique. To produce a topographic image, the probe is approached to the sample in predefined step sizes, typically 10–100 nm.[50] After each step, a short current pulse at the pipette tip determines R_{AC}. When the calculated resistance reaches a predetermined value (typically 3%–10% of R_{AC} measured when the pipette is far from the surface), the Z position of the probe is recorded and the pipette is retracted. The pipette is then moved laterally to a new position and the process is repeated. One disadvantage of backstep mode is that high-resolution images are relatively time consuming to record. For instance, a 15 μm × 15 μm area of a hippocampal cell imaged with a 250-nm lateral step size and 2 nA pulse height required 30 min to complete.[50] Larger lateral step sizes result in less time-consuming, but lower resolution images.

To enable backstep mode for faster imaging of dynamic cellular surfaces, "floating" backstep mode was developed.[52] In this configuration, different scan parameters are used for high aspect ratio areas of the sample relative to flat regions. First, a low-resolution image of the sample is generated to determine surface topography. High aspect ratio areas are identified and imaged with additional lateral points and larger, more time-consuming backsteps than the flat regions of the sample. This process reduces the total number of steps the probe moves during the imaging process. As a result, a 30 μm × 30 μm area of an oligodendrocyte was imaged with 100 nm vertical steps and 500 nm lateral movements in ~20 min. With faster imaging capabilities, SICM has been used to monitor cellular movements and volume changes with 500 nm lateral resolution and 50 nm vertical resolution.[52]

A technique similar to picking and backstep modes is hopping mode.[48] Like picking mode, the hopping mode probe approaches the sample surface until a predefined ion current value is reached, typically 0.25%–1% of the reference current recorded prior to approach. Hopping mode overcomes ion current drift through continuous update of the reference current each time the probe is moved laterally. Similar to floating backstep mode, a low-resolution image is first produced to roughly estimate the complexity and roughness of the surface. High aspect ratio areas of interest are then scanned with a greater number of imaging points and a slower scan time. Hopping mode was first applied to image biological samples with complex surface topography. For instance, mechanosensitive sterocilia of auditory hair cells and hippocampal neurons were imaged with lateral resolution better than 20 nm.[48] However, high-resolution imaging results in slow image acquisition.

Standing approach mode was developed simultaneously by Matsue and coworkers.[49] A minor difference from hopping mode is the establishment of a new set point every time the probe is laterally moved to a new position. To image a surface, the probe is first positioned far from the substrate and a set point ion current is established. The pipette is then approached to the sample with a velocity of 20–100 nm/ms until the predetermined ion current is reached, followed by pipette retraction and lateral movement to a new area. Like hopping mode, a disadvantage of standing approach mode is long image acquisition time. A modification to overcome this limitation uses a field-programmable gate array (FPGA) board, which contains a matrix of reconfigurable gate array logic circuitry that enables an algorithm to control the speed of pipette approach as well as to reduce the distance the pipette is retracted. The FPGA board reduced the scan time of a 100 μm × 100 μm area with 16,384 data points to 7.5 min, compared to ~16 min for unmodified standing approach mode.[49]

2.4.4 FEEDBACK MODE IMPROVEMENTS

The most recent mode to have been developed is fast scanning ion conductance microscopy (FSICM).[55] This configuration is used for fast image acquisition of relatively flat sample features on the same order of magnitude as the pipette tip. Like standing approach mode, an FPGA board is used to enable fast calculation speed. For FSICM, however, a unique ion current feedback calculation process is used. In contrast to typical continuous feedback modes, where a single set point is specified for D_{ps} maintenance, two set points that represent a lower and upper D_{ps} limit are used. If either set point is reached during a single line scan, the pipette position is adjusted for the subsequent line. This technique is useful for flat surfaces because the Z position of the probe cannot dynamically adjust to account for feature height changes, which is the basis for fast image acquisition. An FSICM image is processed with surface topography reconstruction. Within each acquired data point is a pixel, which corresponds to an X,Y location, an ion current measurement, and a topographically inaccurate Z position. The incorrect Z position value can be corrected by correlation of D_{ps} with ion current signal via an approach curve. Addition or subtraction of the calculated D_{ps} value from the Z coordinate of each pixel produces true surface topography measurement. FSICM was compared to hopping mode to assess the differences in scan time and image quality between the two techniques. For hopping mode, the highest practical image resolution is 512 × 512 pixels which can take 5 h to complete. In comparison, FSICM can generate the first 600 lines of a 1024 pixel image in ~10 min. Image quality of these techniques was directly compared with

Xenopus A6 cell images. For hopping mode, a 40 μm × 40 μm area with 128 pixel resolution was imaged and for FSICM, 600 lines of the same area was scanned with 1024 pixel resolution. Hopping mode required 25 min to generate an image of the Xenopus A6 cells while FSICM took only 10 min. More surface detail was observed in the FSICM image because the number of pixels was greater than with hopping mode; however, the rapid probe motion disturbed fragile surface features and created a blurred effect.[55] A limitation of FSICM is the quasi-constant height nature of the probe, which can result in the imaging of small features at nonoptimal D_{ps}.

A practical application of FSICM is the imaging of single nanoparticles in live cells.[56] FSICM was combined with fluorescence imaging to track the movement of 200 nm carboxyl-modified latex particles in immortalized human alveolar epithelial type 1 (AT1)-like cells. The AT1-like cells were transfected with actin-binding protein—green fluorescent protein or clathrin light chain-enhanced green fluorescent protein. These fluorescent modifications allowed the molecular identity of the membrane–nanoparticle interactions to be identified, as well. Combined topographic-fluorescence images were acquired at a rate of 15 frames/second, and complete particle invagination was observed in as little as 14 min.[56] When combined with fluorescence imaging, FSICM is a powerful tool capable of providing topographic images with high temporal resolution, as well as information related to co-localization of proteins.

A recent advance in feedback mode operation, termed bias-modulated SICM (BM-SICM), utilizes modulation of the bias applied between reference and pipette electrodes.[57] Amplitude and phase of the oscillating ion current can both be used as reliable feedback signals. BM-SICM has been developed as an alternative to AC feedback mode, where the oscillating tip can cause convective fluid movement over the sample. Additionally, the feedback response time in AC feedback mode is limited by the probe oscillation frequency. The use of amplitude or phase as a feedback signal produced identical topographic images of gold bands over glass when directly compared to images obtained with AC feedback mode.[57] An important implication of BM-SICM is the possibility for impedance measurements to be acquired, which will extend the applicability of SICM. Table 2.1 summarizes the characteristics of advanced feedback modes.

2.5 RESOLUTION

Scanning probe microscopy image resolution is a complex issue that can be difficult to define. Many factors are involved in high-resolution imaging which broadly fall into the categories of D_{ps}, probe geometry, and feedback.[58] Probe–surface distance is an important consideration because SPM image formation is dependent upon detection of the interaction between probe and sample. In general, a smaller D_{ps} can more accurately measure surface properties. A complication to this rule is when the probe–surface interaction is limited in lateral length scale. For instance, scanning near-field acoustic microscopy cannot achieve atomic resolution, as the acoustic interactions that govern this technique do not vary on the Ångström scale.[58] On the other hand, D_{ps} cannot be reduced beyond a finite limit for some techniques. For instance, when the probe and surface are both conductive in SECM experiments and D_{ps} is sufficiently small (less than 30–50 nm), tunneling current is detected instead

TABLE 2.1

Comparison between Specialized Feedback Modes Developed for SICM

Mode	Substrate or Sample	Advantages	Disadvantages	Scan Speed	Reference
Picking	Conductive or insulating surface in redox mediated solution	Decouples topography and surface conductance measurements	Convective processes unknown	70 μm/s	44
Hopping	Complex surface	Stable, insensitive to ion current variation, can image complicated features	Slow scan time for high resolution images	25 min for 40 μm^2, at 128 $\times$ 128 pixels; 5 h for 40 μm^2 at 512 $\times$ 512 pixels	45
Standing	Complex surface	Fast image acquisition of complex samples	Requires specialized electronics	7.5 min for 100 μm^2 with 16,384 data points	46
Backstep	Cultured cells	Overcomes drift, can access overhanging surfaces	Low resolution; slow	30 min for 15 μm^2 with 500 nm vertical steps and 250 nm lateral steps	47
Floating Backstep	Cultured cell movements and volume changes	Faster image acquisition than backstep mode	Increased step size doubles scan time	20 min for 30 μm^2 with 100 nm vertical steps and 500 nm lateral steps	49
Fast Scan	Flat surface	Fast image acquisition, can access many imaging points	Decreased scan time decreases resolution; difficult image processing; limited Z range	409.6 ms/line at 1024 $\times$ 600 pixels	52

of electron transfer.[59] Another consideration is that D_{ps} must be maintained at a safe imaging height, such that the probe is protected from contact with the sample. While small D_{ps} is important for SPM imaging, many factors must be balanced to achieve optimum detection of probe–surface interaction.

Probe size and geometry are essential considerations for resolution. Probe dimensions are a crucial factor to determine the feature size that can be accurately measured with a given technique. Ideally, a probe should be significantly smaller than the feature of interest because a miniaturized probe reduces probe–surface interaction averaging, which can convolute images and decrease resolution.[58] Probe geometry is equally important, especially when measurement quantification relies on equations which assume probe shape. However, limitations of small probes include probe fabrication and system sensitivity. Nanoscale probes can be difficult to produce and often require advanced instrumentation or technical knowledge. Although smaller probes enable greater sensitivity for SPM measurements, not all systems are presently capable of robust detection and control at reduced scales.

The SPM feedback mechanism, if applicable, has important implications for resolution. When D_{ps} is controlled by feedback, instability of the probe must be reduced to ensure proper tip position is measured. When probe position is not precisely maintained, image convolution can occur. Quantification of data can also be difficult, when D_{ps} is not highly controlled. Inherent differences among various feedback modes for a given system affect image resolution, thus care must be taken to select the mechanism most suitable for the chosen sample.

Multiple factors contribute to the overall spatial resolution of an SPM technique, which is the ability to distinguish features within an image. A type of spatial resolution is lateral resolution, defined as the smallest distance between two features in which both features are differentiated. Lateral resolution can be characterized by the Rayleigh criterion, which is a condition that must be met for the limit of lateral resolution to be observed. This is easily visualized when considering the case commonly encountered in optical microscopy and the Rayleigh criterion, a model that considers point source diffraction patterns generated when light passes through a lens.[60] According to the Rayleigh criterion, the lateral resolution limit is met when the maximum intensity of a point source overlaps the minimum intensity of a second point source. The Rayleigh criterion can be applied to SPM techniques when intensity profiles are replaced with topographic line profiles taken over feature.[61] Examples of two features which are resolved, unresolved, or at the limit of lateral resolution are shown in Figure 2.5.

Lateral resolution in SICM has been examined both theoretically and experimentally.[62,63] An experimental investigation was first performed by Shevchuck et al. in

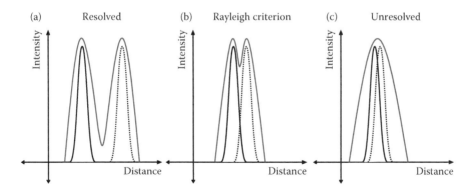

FIGURE 2.5 Resolved and unresolved line profiles based on Lord Rayleigh's criterion, which states that the limit of resolution occurs when the maxima of one peak overlaps the first minima of a second peak. (a) Two resolved peaks. (b) Two peaks which meet Rayleigh's criteria. (c) Two unresolved peaks. (With kind permission from Springer Science + Business Media Rayleigh, L., *The London, Edinburgh, and Dublin Philosophical Magazine and Journal of Science* **1879**, *8*, 261–274. Copyright 2007 Springer.)

2006.[62] The S-layer of a *Bacillus sphaericus* cell, which contains proteins spaced 13.1 nm apart with high periodicity, was imaged with a 13-nm inner diameter pipette. Raw topography images were processed with a 2D fast Fourier transform, and spatial frequencies that corresponded to 3, 6, and 12 nm resolutions were examined. Between the spatial frequencies of 3 and 6 nm, distinct features were visible, which led to the conclusion that the best SICM lateral resolution that can be achieved is 3–6 nm. Later, a model was developed with finite element method (FEM) simulations.[63] Two spherical particles were simulated and spaced apart as a function of the inner radius (r_i) of a hypothetical pipette, from $6r_i$ to r_i. Simulated topographic images were produced and line profiles taken over the images revealed the particles spaced $4r_i$ apart were fully resolved, and those spaced $2r_i$ were unresolved. A grayscale plot of each simulated image and line profile as a function of particle distance was produced and lateral resolution was determined to be ~3x the inner radius of the pipette.[63]

Besides lateral resolution, FEM simulations have been used to investigate other characteristics of SICM resolution. The effect of surface topography on image formation was examined by Unwin and coworkers.[38] In a FEM model, pits of various radii and depths were examined with simulated approach curves. The relationship between probe radius, pit radius, and pit depth was evaluated. For pits significantly smaller than the probe radius (e.g., 0.73 times), changes in surface topography were not detected by the probe. For pits significantly larger than (e.g., 5.74 times) the probe radius, pit depths were accurately measured. The implication of this study is that probe size must be sufficiently smaller than feature size for true feature dimensions to be reproduced. This result was verified by another simulation, performed by Schaeffer and coworkers, who used spherical particles of various radii to investigate the relationship between feature size and probe size.[63] Simulated line profiles of the particles indicated particle dimensions were most accurately reproduced when particle diameter was at least $6r_i$ of the theoretical pipette. These studies confirm probe size is an important experimental consideration.

The effect of pipette geometry on image resolution was also examined with FEM simulations. Typical geometric parameters are cone angle, inner radius, outer radius, and RG value. The effect of cone angle on current sensitivity was simulated and revealed to be an important factor for the current-distance response of the probe.[38] Theoretical approach curves were generated for a pipette with a constant RG value of 1.1 and cone angles of 3°, 5°, 10°, and 15°. The approach curves indicated ion current sharply decreased with D_{ps} as cone angle increased. This phenomenon was explained with consideration of tip resistance. When the cone angle is small, the resistance of the pipette tip is large and contributes predominantly to overall resistance. When the cone angle is large, resistance measured has a greater contribution from R_{AC}. Thus, at smaller D_{ps}, greater current sensitivity is achieved with larger cone angles. A similar effect was observed with RG value. Simulations were performed with a pipette of cone angle 15° and RG values of 1.01, 1.1, 2, and 10. Current sensitivity increased as RG increased.[38] However, this study was first performed in DC feedback mode, and the effect of RG on current sensitivity was found to be complicated when tip position was modulated due to variations in potential along the tip annulus. In AC mode, pipettes were still more sensitive to ion current changes at larger RG values, but higher D_{ps} was necessary which resulted in decreased resolution. Knowledge of pipette parameters and their effects on resolution are important for optimum image quality.

The effect or D_{ps} on resolution was also examined both theoretically and experimentally. In an FEM study, spherical particles of various sizes were simulated and imaged at different D_{ps}.[63] In every instance, as D_{ps} decreased, particle dimensions such as height and width were more accurately measured. The same result was observed experimentally when raised bands on a silicon wafer were imaged at two different set points.[38] Topographic images of the two situations qualitatively showed better resolution achieved with smaller D_{ps} values. Quantitatively, overlaid line scans of both images indicate smaller D_{ps} more closely approximated the true dimensions of the raised band feature than when the probe was farther from the surface. The importance of imaging with small D_{ps} is underscored in these studies.

Instabilities in the feedback system are detrimental to resolution and include poor piezoelectric positioner response, cross talk, and noise. When high aspect ratio features are imaged, the pipette may retract too quickly in the Z direction which can result in lag of the feedback loop. Cross talk can be observed between topographic and current images of a substrate with both conductive and nonconductive features are generated.[8] As the probe is scanned across a sample surface that transitions from nonconductive to conductive, the feedback mechanism assumes the tip is closer to the surface than the set point, which results in a decreased ion current signal. The opposite occurs when the probe is scanned from conductive to nonconductive regions. These effects can be minimized with decreased scan speed and optimized scan parameters.

Modification of AC feedback mode was performed by Klenerman and coworkers,[43] who evaluated resolution when a lock-in algorithm was used to recover ion current signal over a noisy background on a convoluted surface. The improved method was developed to overcome AC feedback mode limitations on precise topographic imaging, when a complex surface is scanned. The need for an improved method was demonstrated with a simulated line scan over complicated surface features, which showed convolution in the ion current signal caused by slopes in the surface topography that resulted in noise and incorrect D_{ps} measurement. These surface components were removed with a lock-in amplifier and mathematical algorithm. A PDMS grid was imaged with both feedback techniques and superior resolution was recorded with the improved method. Smoother line scans, stable imaging, and more reliable depth profiling were observed for this case.[43]

2.6   HYBRID TECHNIQUES

2.6.1   Electrochemically Insensitive Techniques

Scanning ion conductance microscopy has been incorporated into a variety of non-electrochemical techniques to impart topographic imaging capabilities or improved probe control. Integration of SICM with confocal microscopy has enabled the development of an SPM technique called scanning surface confocal microscopy

(SSCM)[64–68] that combines fluorescence and noncontact topographic imaging for high-resolution, quantitative biological studies. A typical SICM setup can be placed atop an inverted confocal microscope, such that the SICM probe is positioned above the sample and the excitation beam for fluorescence imaging is positioned below. As the sample is scanned relative to the probe and beam, simultaneous topographic and fluorescence images are produced. This technique was first used to study particle entry into cellular membranes with virus-like particles invaginated into the cellular surface of monkey COS 7 cells.[65] The high resolution achieved with SSCM enabled the lateral movement and penetration characteristics of single particle entry to be measured. The simple integration of SICM into other instrumental setups makes SICM a valuable modification for optical techniques.

More commonly, SICM has been combined with existing SPM techniques, such as scanning near field optical microscopy (SNOM)[44,69–74] which makes use of a near-field light source with an output aperture of sub-wavelength dimensions to scan samples for high-resolution topographic and optical imaging. The advantage of SICM integration with SNOM is to enable precise probe control. Scanning ion conductance microscopy and SNOM probes are similar in design, as both are tapered to a nanoscale aperture tip. However, SNOM probes typically are pulled optical fibers made conductive through metal deposition. The hybrid technique was first used to image rabbit cardiac myocytes.[71] A continuous wave laser was coupled to a multimode fiber in an aluminum-coated SICM probe. High-resolution topographic and optical images of the myocytes both showed sarcolemic striation patterns on the cells, which corresponded to observations made with electron microscopy. These results represented the first time simultaneous topographic and optical images were produced in the near field. A later integration of the two methods used SICM as a novel light source for SNOM. Reaction of a fluorophore at the nanopipette tip with a ligand in the electrolyte bath produced a suitable amount of light for fluorescence excitation. The nanopipette probe was filled with calcium-free fluo-3 solution and the bath electrolyte contained 2.5 mM Ca^{2+}. Contact between nanopipette and bath solutions generated the fluorescent complex fluo-3-Ca^{2+} at the pipette tip, which was then excited with an Ar^+ laser to produce a continuous light source for SNOM imaging. Sperm flagella and Xenopus A6 kidney epithelial cells were imaged in physiological buffer with ~190 nm resolution.[71]

Scanning ion conductance microscopy has also been paired with atomic force microscopy (AFM).[45,75] For an SICM-AFM configuration, a bent nanopipette was used as the cantilever probe. Probe–surface distance was recorded by laser deflection off the back of the pipette, and ion current was collected at the nanopipette opening.[45] With this modification, conductive pathways of synthetic polycarbonate membranes were recorded.[45] Other hybridizations of AFM and SICM have been used to identify the mechanosensitive properties of living cells[75–77] and guide neuronal growth cone development.[78]

2.6.2 Electrochemical Measurements

Recently, methods that combine SECM with SICM to enable electrochemical sensing at the nanopipette have been of special interest. The advantage of SICM-SECM hybridization is the creation of a multifunctional tool, which overcomes key limitations of each individual technique. Scanning ion conductance microscopy is "chemically blind," while SECM lacks precise probe control. The two methods can be combined with fabrication of a dual probe capable of simultaneous ion current and electrochemical signal detection. A number of creative dual probe fabrication strategies have been described.[16,18–21]

Two general dual probe fabrication techniques for simultaneous ion current and electron transfer measurements have been reported. In the single-barrel dual probe (SBDP) approach,[18–20] a nanopipette is first coated with a conductive material, such as gold, to create a ring[19–20] or crescent-shaped electrode[18] around the pipette tip. An insulating layer is then deposited around the entire probe and a small electrode area at the pipette tip is revealed through focused-ion beam (FIB) milling, masking, or mechanical polishing. A schematic of the SBDP is shown in Figure 2.1b. In Figure 2.1e, an electron micrograph of a dual probe with a ring electrode geometry is shown.

The SBDP fabrication technique was used to create a gold ultramicroelectrode (UME)/nanopipette probe.[20] Gold was first deposited onto one half of a glass pipette, followed by atomic layer deposition of aluminum oxide over the entire probe. A gold electrode with a 294-nm radius and pipette with 100 nm inner diameter was exposed with FIB milling. Approach curves generated when the hybrid probe was positioned over conductive and insulating surfaces in a solution of 10 mM $Ru(NH_3)_6Cl_3$, and 100 mM KNO_3 showed that the UME behaved in a manner consistent with typical SECM experiments. Ion current measurements at the nanopipette were independent of the UME, as indicated by approach curves taken over conductive and nonconductive surfaces. Topographic and electrochemical image resolution of the SBDP was examined with a resolution standard that contained alternating 400 nm wide gold stripes and glass trenches spaced ~2.6 μm apart. The trenches were successfully resolved in topographic images. Redox current images showed positive feedback over conductive areas and negative feedback over the insulating glass portions of the sample, which is consistent with a typical SECM experiment. The subsequent topographic and faradaic images are shown in Figure 2.6. The spatial resolution attained with the dual probe was further characterized with 180 nm wide trenches separated by 875 nm. These features were successfully resolved, but with poorer image quality than the previously imaged standard. The electrode was tested in substrate generation/tip collection (SG/TC) mode by imaging a 30-nm thick, 10-μm wide gold electrode in 10 nM $K_3Fe(CN)_6$ and 100 mM KCl. From topographic and electrochemical images, the electrode dimensions were well-resolved and enhanced oxidation current was generated over the gold electrode, as would be expected for SG/TC mode. These results show the dual probe can independently generate both topographic and ion current images with high resolution and localized recycling of UME-generated redox species over a conductive surface can be achieved.

An SBDP fabrication strategy was also reported by Takahashi et al.[19] with a 330-nm inner radius Pt or Au nanoring electrode fabricated around a 220-nm inner radius nanopipette. Unlike the originally reported SBDP, the probe reported here was used in hopping mode and with ion current feedback for electrochemical imaging. Probe resolution was characterized for electrochemical imaging with ion current feedback

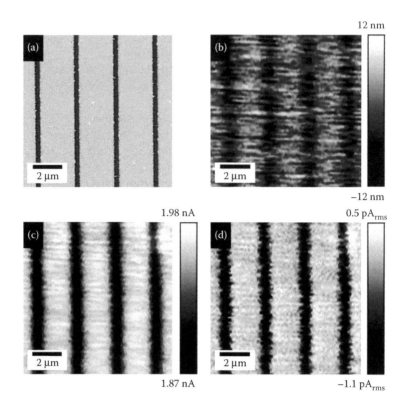

FIGURE 2.6   Scanning electrochemical micrograph, topographic, and redox current images of 400 nm trenches FIB milled onto gold-on-glass substrate. (a) Scanning electrochemical micrograph. (b) SECM-SICM topography. (c) SECM-SICM DC redox current. (d) SECM-SICM AC redox current. (Reprinted with permission from Comstock, D. J. et al., Integrated ultramicroelectrode–nanopipette probe for concurrent scanning electrochemical microscopy and scanning ion conductance microscopy. *Analytical Chemistry* **2010**, *82*, 1270–1276. Copyright 2010 American Chemical Society.)

over a Pt interdigitated array (IDA) electrode in 0.50 mM $FcCH_2OH$ and 0.1 M KCl. The IDA electrode height estimated from the topographic image was 100 nm and in good agreement with dimensions determined by conventional AFM. Topographic and electrochemical images of the electrode showed simultaneous SICM and SECM measurements could be made at sub-micron resolution.

The hybrid SBDP was used to measure enzyme activity of patterned glucose oxidase (GOD) (shown in Figure 2.7) and horse radish peroxidase (HRP) monolayers with SG/TC feedback mode while D_{ps} was maintained with ion current feedback. Simultaneous topographic and electrochemical images of enzyme activity in GOD and HRP patterns were produced with high resolution, which is notable because previous attempts to image enzyme arrays with SECM have been limited by low image resolution.[79–81] Incorporation of ion current feedback into the SECM system is especially advantageous for resolution. The influence of D_{ps} on SG/TC imaging was investigated when ion current feedback was used to position the probe 100 and

Topography   Electrochemical image

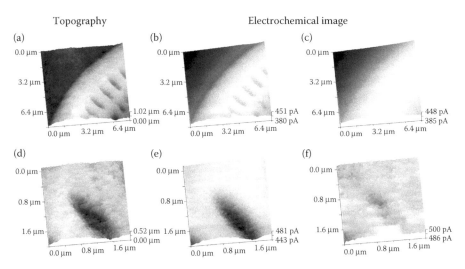

FIGURE 2.7 Topographic and electrochemical images of glucose oxidase (GOD) substrate imaged in 20 mM glucose in 0.50 mM FcCH$_2$OH and 0.1 M KCl. (a) SECM-SICM topographic image, scan size 8 μm × 8 μm. (b) Electrochemical image, scan size 8 μm × 8 μm and probe–surface distance 100 nm. (c) Electrochemical image, scan size 8 μm × 8 μm and probe–surface distance 600 nm. (d) SECM-SICM topographic image, scan size 2 μm × 2 μm, (e) Electrochemical image, scan size 2 μm × 2 μm and probe–surface distance 100 nm. (f) Electrochemical image, scan size 2 μm × 2 μm and probe–surface distance 600 nm. (Reprinted with permission from Takahashi, Y. et al., Simultaneous noncontact topography and electrochemical imaging by SECM/SICM featuring ion current feedback regulation. *Journal of the American Chemical Society* **2010**, *132*, 10118–10126. Copyright 2010 American Chemical Society.)

600 nm above the GOD monolayer pattern. At 100 nm D_{ps}, fine, cave-like structures were observed in the electrochemical image, but not when the probe was held 600 nm above the surface.[17] Precise D_ps control also enabled SECM-SICM to generate electrochemical images of tight junctions, varicosities, and ridge-like structures of superior ganglion and Xenopus A6 kidney epithelial cells. Complex electrochemical images were previously difficult to reliably obtain.[19]

The SBDP also enabled heterogeneous cell permeability to be examined which SECM was also previously unable to characterize. Faradaic current for oxygen, potassium ferrocyanide (K$_4$[Fe(CN)$_6$]), and ferrocene methanol (FcCH$_2$OH) on a cardiac myocyte was monitored. For oxygen, reduction current magnitude at the cell membrane was large compared to the current observed over the petri dish, which indicated high membrane permeability for oxygen. For K$_4$[Fe(CN)$_6$], a hydrophobic mediator, the oxidation current at the cellular, and petri dish surfaces was the same, which showed diffusion of K$_4$[Fe(CN)$_6$] was blocked. The redox mediator FcCH$_2$OH was permeable in the cell membrane, as oxidation current measured at the membrane was enhanced compared to the culture dish surface. The observation that oxygen and FcCH$_2$OH were permeable in the cell but Fe(CN)$_6$ was not signifies heterogeneous cell permeability was detected.[19]

The SBDP fabrication technique was also applied for the measurement of redox probe diffusion through porous membranes.[18] A gold electrode was fabricated around the nanopipette in a "half-moon" geometry through thermal deposition and subsequently insulated with parylene C. A small region at the tip of the probe was masked with PDMS prior to insulation to expose the electrode upon mask removal. Nanopipettes with outer diameters less than 500 nm and Au electrodes ~100 nm in diameter were produced. Electrochemical images of redox probe diffusion through a porous membrane are shown in Figure 2.8. A membrane with a 900-nm diameter pore was placed in a perfusion cell with 100 mM KCl in the top chamber and 100 mM KCl with anionic (ferricyanide) or cationic (ruthenium hexamine) redox probe in the bottom chamber. Topographic, ion current, and faradaic images were generated simultaneously when ions and redox molecules were driven through the pore via a concentration gradient and transmembrane potential applied to the working electrode. The presence of the redox mediator increased the ion current response measured with SICM due to changes in solution conductivity in the vicinity of the pore. For instance, when ferricyanide migrated through the perfusion cell under negative transmembrane potentials, the ion current value recorded was greater than when no redox probe was present in solution. As the transmembrane potential increased, ion current signal increased as well. The SBDP was also shown to have on/ off electrochemical detection capability for oppositely charged probes. For instance, when a negative transmembrane potential was applied when ferricyanide was present in the bottom cell of a perfusion chamber, electrochemical current was imaged in the vicinity of the pore. However, when a positive transmembrane potential was

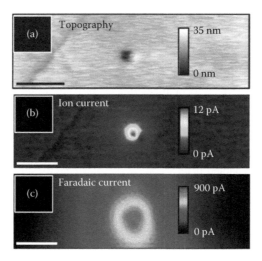

FIGURE 2.8 Topographic, ion current, and faradaic current images of redox probe diffusion through a single, 900-nm diameter nanopore. Scale bars are 2 μm. (a) SECM-SICM topography image. (b) Ion current image. (c) Faradaic current image. (Morris, C. A.; Chen, C.-C.; Baker, L. A., Transport of redox probes through single pores measured by scanning electrochemical-scanning ion conductance microscopy (SECM-SICM). *Analyst* **2012**, *137*, 2933–2938. Reproduced by permission of The Royal Society of Chemistry.)

applied, no electrochemical signal was detected. The same principle was observed for ruthenium hexamine, except that electrochemical signal was detected under positive transmembrane potential and no change was observed otherwise.[18]

A second strategy for dual electrode fabrication uses a double-barrel pipette to produce double-barrel carbon nanoprobes (DBCNP) for SECM-SICM characterization and voltage-driven delivery.[21] A caveat to pyrolysis inside of the double-barrel probe is the lack of a well-defined probe geometry, which can result in difficulty in quantitative treatment of the data. A schematic of a dual-barrel probe is shown in Figure 2.1c and a scanning electron micrograph of a DBCNP is shown in Figure 2.1f. To fabricate the electrochemically sensitive portion of the probe, one barrel of the pipette was filled with butane gas and pyrolyzed to form a conductive carbon electrode. The other pipette barrel was used for ion current detection. The DBCNP design strategy has advantages over the SBDP method. For instance, the double-barrel probes have better insulation than coated counterparts, fabrication does not require special instrumentation and probe radii as small as 10 nm have been reported.[21] Topographic and electrochemical resolution of the probe was investigated with polyethylene terephthalate (PET) membranes and Pt IDAs. Concurrent high-resolution topographic and electrochemical images were generated with features as small as ~100 nm faithfully reproduced. The capability of the probe for biological imaging was examined with live cells. Topographic images generated with DBCNPs of differentiated PC12 cells were compared to the same images taken with a single-barrel SICM nanopipette. Similar resolution was observed in both cases which indicated integration of a UME next to a pipette did not alter ion current detection capability in the dual probe. A unique application of the DBCNP is localized voltage-driven delivery of ions through the SICM barrel onto a substrate. The probe's ability to precisely deposit ions was tested with polarizable PC12 cells. Stimulation of the cells by K^+ deposition when the carbon electrode was biased to +1000 mV resulted in depolarization and dopamine release, which was subsequently imaged electrochemically. The high-resolution and precise probe control of DBCNPs make voltage-driven delivery a potentially useful biological application.

A method which elegantly utilizes one pipette probe to record topography, conductance, and electrochemical measurements, termed scanning electrochemical cell microscopy (SECCM), has been described.[82] Here, a dual-barrel pipette, where each barrel contains an Ag/AgCl wire electrode and is filled with electrolyte solution, is used for imaging. A meniscus is formed at the probe tip and an applied bias enables ion current flow across the meniscus. When the meniscus is in contact with a semi-conductive or conductive surface, the probe is oscillated and ion current feedback is used to maintain a set probe–surface distance. An advantage of SECCM over SICM or SECM is that samples do not have to be immersed in solution, which allows both dry and wet samples to be imaged. Imaging under dry conditions is useful for studies where changes in the electrode surface may occur if samples are biased in liquid for long periods of time. SECCM has been used for a number of interesting studies,[83] such as simultaneously recording conductance, electrochemical, and topographic images of electrocatalysis of platinum nanoparticles on a single carbon nanotube.[79] Notably, correlation of surface topography with heterogeneous electron transfer rates and kinetics of electrode surfaces have been reported with SECCM.[80,81,84–87] Correlation of topographic and electrochemical images of platinum electrodes in ionic liquids[88] have also been achieved via SECCM. A more comprehensive review of the technique is available.[89]

2.7 ION CURRENT AS SIGNAL

2.7.1 Conductive Pathways in Nanoporous Membranes

The ability to measure conductive pathways has made SICM a useful tool for ion transport studies. Since the original report[4] where SICM was used to produce topographic and ion current images of polymer membranes with micron-sized pores, the ion current sensing capability of SICM has been extended to a variety of substrates from lithium ion batteries to live cells. Instrument development has further enabled SICM to measure ion current transport through nanoscale features,[11–14] such as single nanopores[11] or multiple nanopores with heterogeneous conductance.[12] To measure transport through a single nanostructure, a three-electrode system was developed in which a working electrode was placed beneath a porous membrane in a perfusion cell and a transmembrane potential was applied to drive ions through the pore.[11] A scanning electron micrograph of a single pore is shown in Figure 2.9a. The absolute pipette current versus transmembrane potential was recorded for five probe–surface distances from 610 nm above to 2 µm inside the pore. As D_{ps} decreased ion current magnitude increased due to a change in R_{AC} and enhanced interaction between the nanopore and the charged walls of the pipette. Current–voltage responses were also dependent on lateral displacement of the pipette from the pore center. When the probe was positioned 330 nm away from the feature, the current–voltage response significantly decreased. As with vertical probe position, this phenomenon was due to the change in R_{AC}.[11]

For heterogeneous transport studies a Pt counter electrode was added to the three-electrode system to help prevent potential fluctuations on the reference electrode for stable ion current measurement.[12] These studies were performed with conical pores, which taper at one end and display ICR due to the asymmetric geometry of the pore. When negative transmembrane potentials were applied to the base of the conical pore, larger ion current responses were observed than when a positive transmembrane potential was applied. Variations in the ion current response of six conical pores were investigated as a function of transmembrane potential. Results from this study are shown in Figure 2.9b,c. Ion current rectification ratios (ICRRs) for each pore were calculated. Briefly, ICRR can be described with Equation 2.7, where Δi_{+v} is the change in current with respect to positive applied potential and Δi_{-v} is the change in current with respect to negative applied potential.

$$ICRR = \frac{\Delta i_{-V}}{\Delta i_{+V}} \qquad (2.7)$$

Rectification occurs when ICRR is greater than 1. ICRRs between 1.1 and 66.0 were observed for the six features. The variation in ion current signals suggested the

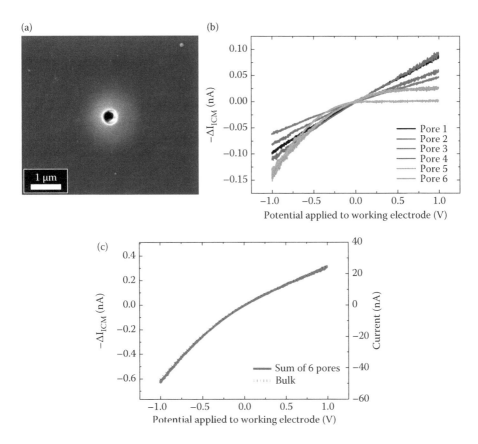

FIGURE 2.9 Scanning electron micrograph of pore used for single nanopore studies and current–voltage curves of individual pores. A current–voltage curve is generated over six individual conical pores. Probe–surface distance is maintained at 180 nm above the pore. Each pore is shown to have a different rectified response, which agreed with the aggregate response of the entire membrane. (a) Single nanopore. (b) Individual current–voltage response of six pores. (c) Aggregate response. (Panel a reprinted with permission from Chen, C.-C.; Zhou, Y.; Baker, L. A., Single-nanopore investigations with ion conductance microscopy. *ACS Nano* **2011**, *5*, 8404–8411. Copyright 2011 American Chemical Society; Panels b and c reprinted with permission from Zhou, Y.; Chen, C.-C.; Baker, L. A., Heterogeneity of multiple-pore membranes investigated with ion conductance microscopy. *Analytical Chemistry* **2012**, *84*, 3003–3009. Copyright 2012 American Chemical Society.)

ion transport properties of each pore were independent of one another. Correlation of the individual current–voltage responses with the bulk response of the membrane, measured with conventional SICM, showed good agreement between the two methods.[12]

Scanning ion conductance microscopy was further modified for the auxiliary detection of potentiometric signal with a technique called potentiometric-scanning ion conductance microscopy (P-SICM).[17] The addition of a fifth electrode to the four-electrode system described for heterogeneous ion transport enabled the additional measurement. The extra electrode was placed in one barrel of a theta pipette, while the other barrel was used for ion current detection. This system has been used for biological studies, as potentiometric detection is more sensitive than ion current measurement and enables smaller transmembrane potentials to be applied to cellular membranes which improves prolonged cell viability.[17] Potentiometric SICM was used to distinguish para- and transcellular transport pathways across epithelial cell monolayers.[17] An epithelial monolayer consists of cell bodies, where transcellular pathways are located, and cell junctions, the area between cells where paracellular pathways can be found. The P-SICM probe was held above a cell body or cell junction at high (12.5 μm) or low (0.2 μm) D_{ps}, while the working electrode was scanned ±50 mV. Different magnitudes of potential deflection at each position suggested the presence of conductive pathways. Additional conductance measurements revealed the cell body and cell junction had distinct relative conductance values, as shown in Figure 2.10. This result is important because local paracellular conductance had not previously been measured with sub-micron resolution.

A five-electrode system was used with an SBDP with a gold electrode to measure potentiometric response correlated to local pH change through a porous membrane.[16] The same SBDP fabrication method reported for redox probe diffusion[18] was used, but electropolymerization of aniline to polyaniline (PANi) film on the Au portion of the probe enabled pH measurement. A porous polyimide membrane was mounted in

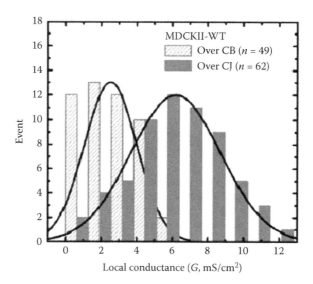

FIGURE 2.10 Apparent conductance differences for transcellular and paracellular pathways. When the P-SICM probe is positioned over the cell body or cell junction of an epithelial cell monolayer, different apparent conductances are observed, which correlate to transcellular and paracellular pathways, respectively. (Reprinted with permission from Chen, C.-C. et al., Scanning ion conductance microscopy measurement of paracellular channel conductance in tight junctions. *Analytical Chemistry* **2013**, *85*, 3621–3628. Copyright 2013 American Chemical Society.)

a perfusion cell, and ion current flux was imaged while solution pH was changed. Potentiometric images from this process are shown in Figure 2.11. The SBDP had a dynamic range of pH 2.5–12.0 and a sensitivity of pH 0.02. Lateral resolution was investigated by measuring pH response for two pores spaced 3.7 μm apart. The response between the pores was resolved, and slightly different pH responses were measured, which indicated the independence of the structures. Vertically resolved pH measurements were obtained as well. When the probe was lowered into the pore, a 0.13 pH difference was measured relative to when the probe was positioned directly above the pore.[18] More recent reports by Unwin and coworkers have also described hybrid potential measurements with SICM.[90] Overall, modifications to the original SICM configuration have enabled more sensitive ion current detection as well as additional signal to be recorded.

2.7.2 Li⁺ Ion Electrochemistry

Scanning ion conductance microscopy has been used for *in situ* characterization of Li⁺ battery electrochemistry.[8] Recently, Li⁺ battery technology has been investigated for application in electric vehicles, yet is presently limited by ion current

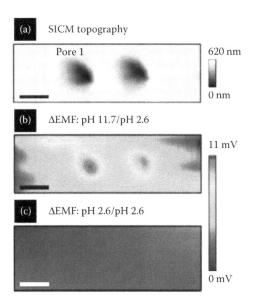

FIGURE 2.11 Topography and EMF images of two laterally spaced pores with and without a pH gradient in a diffusion cell. Scale bars are 2 μm. (a) SICM topography image of two pores in a polyimide membrane. (b) EMF image of local pH difference generated with a pH gradient of 0.1 M KCl, pH 11.7 in the top chamber of a diffusion cell and 0.1 M KCl, pH 2.6 in the bottom chamber. (c) EMF image of local pH difference generated with no pH gradient. (Reprinted with permission from Morris, C. A. et al., Local pH Measurement with scanning ion conductance microscopy. *Journal of the Electrochemical Society* **2013**, *160*, H430–H435. Copyright 2013 The Electrochemical Society.)

inhomogeneity due to degradation of anode and cathode materials during normal operation. Several factors contribute to electrode failure such as poor electrical contact to the current collector and formation of a solid electrolyte interphase (SEI) layer.[8] To investigate a Li+ battery with SICM, special modifications were required. For instance, the typical Ag/AgCl pipette electrode was replaced with a lithiated tin wire, and a piece of Li metal was placed in the bath solution (1 M LiPF$_6$) along with a battery such that the sample could undergo lithiation. A battery electrode was prepared with alternating 60 nm thick tin and copper stripes on a glass substrate and imaged during lithiation. An ion current and topographic image of the sample are shown in Figure 2.12. Upon partial lithiation of the surface, structural changes on the battery electrode were observed. Analysis of topographic and ion current images suggested an SEI layer had grown around the perimeter of the sample. Ion current signal was decreased in the SEI region because a barrier to ion current was formed

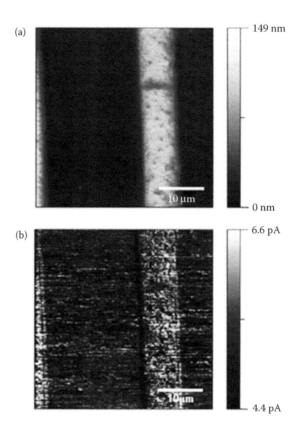

FIGURE 2.12 Direct comparison of electrochemically active and inactive materials typically found in a battery. (a) Topographic image of 60 nm tin stripe deposited onto copper, (b) corresponding ion current image. (Lipson, A. L.; Ginder, R. S.; Hersam, M. C., Nanoscale *In situ* characterization of Li-ion battery electrochemistry via scanning ion conductance microscopy. *Advanced Materials* **2011**, *23*, 5613–5617. Copyright 2011 Wiley-VCH verlas GmbH & Co. KGaA. Reproduced with permission.)

due to reduced Li^+ insertion. These observations suggested the growing solid film could block the tin electrode and result in catalytic decomposition of the electrolyte.[8] The unique application of SICM for battery electrode characterization showcases the versatility of SICM for electrochemical measurements.

2.7.3   ACTUATED CONDUCTING POLYMERS

Scanning ion conductance microscopy was shown to detect redox through measured changes in ion current magnitude as a function of reduction and oxidation of a conducting polymer.[7] A poly(3,4-ethylenedioxythiophene) (PEDOT) film served as the working electrode in an electrochemical cell that was oxidized and reduced during 60-s cycles of ±800 mV applied potential in a bath solution of 0.01 M NaCl. Film topography expanded or contracted due to ion flux into or out of the PEDOT film and solution conductivity changed when oxidizing or reducing potentials were applied. As a result of these changes, ion current signal increased when the film contracted or decreased when the film expanded. Ion current response as a function of redox is shown in Figure 2.13 with areas of interest labeled R1–R5. When +800 mV was applied to the PEDOT film and the polymer was oxidized, a brief film contraction (R1) was observed and the ion current signal increased. Film contraction was followed by sharp expansion of the surface topography (R2) and decreased ion current magnitude that reached a minimum value in this region. Film expansion leveled off over time (R3) and ion current slowly increased. When −800 mV was applied to reduce the polymer, the film rapidly contracted (R4) and the pipette current quickly increased to a maximum value. After initial contraction the surface slowly continued to shrink over time (R5) and ion current magnitude decayed. As the polymer film was repeatedly cycled, ion current maxima and minima became more prominent. Topography changes in the PEDOT film resulted in rapid changes in ion current magnitude because ions near the film surface were quickly expelled out of or passed into the polymer. The sudden change in topography also caused rapid changes in R_{acc} between the film and pipette tip which contributed to the observation of a sharp minimum or maximum. The ion current signal decayed in both regions over time because after initial expansion or contraction, the system equilibrated and ions returned to their original location in the film or solution.[7] Use of SICM for detection of ion current change as a function of redox is a unique solution for the technique's chemical blindness.

2.7.4   INTERFACE OF TWO IMMISCIBLE ELECTROLYTE SOLUTIONS

Scanning ion conductance microscopy was applied to investigate the interface of two immiscible electrolyte solutions (ITIES).[9] Two opposing views describe interfacial structure; one opinion is that solvent dipoles orient to form an ion-free compact layer contained in a molecularly sharp interface. The sharp interface is proposed to separate two back-to-back double layers.[91] The opposite view suggests the interface is composed of a mixed solvent layer that ions of both phases can penetrate.[92] To adequately examine the interface, a technique with high-resolution and small probe are desired. An SECM study performed by Bard and coworkers of a water/nitrobenzene

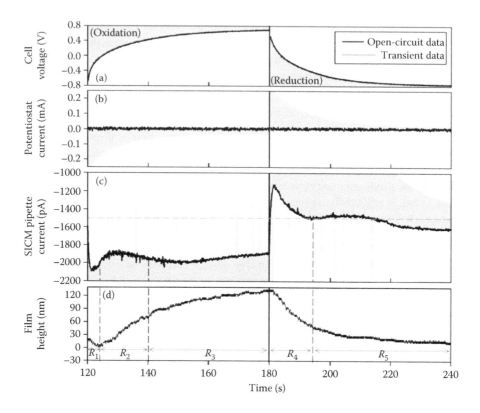

FIGURE 2.13 Potential, current, and topographic profiles due to actuation of a PEDOT film between ±0.8 V. (a) Electrode potential, (b) potentiostat current, (c) pipette current, (d) PEDOT film height changes are shown. Distinct height changes of PEDOT film as a function of reduction and oxidation, denoted R_1–R_5, are identified. (Reprinted with permission from Laslau, C. et al., Measuring the ionic flux of an electrochemically actuated conducting polymer using modified scanning ion conductance microscopy. *Journal of the American Chemical Society* **2011**, *133*, 5748–5751. Copyright 2011 American Chemical Society.)

interface used a UME with a 25-nm Pt tip to estimate the thickness of the interface as 4 nm.[93] Limited probe control and nanoelectrode fabrication strategies hinder SECM, however, for ITIES studies. Shao and coworkers applied SICM for measurement of a water/nitrobenzene interface with a 5-nm inner diameter pipette.[9] An approach curve was generated to measure the current-distance relationship of the pipette to the ITIES, as shown in Figure 2.14. The approach curve can be broken down into three sections: (1) when the pipette approaches the interfacial region, (2) the aqueous phase, and (3) the organic phase. Tip current increased as the pipette approached the interface due to ITIES-hindered ion diffusion and increased cation concentration near the pipette tip. Upon contact with the interface, pipette current instantaneously increased because of interaction with the micrometer-sized diffusion field. Analysis of a cyclic voltammogram taken at this interface suggested the current change was dominated by ion transfer. An estimate of the interfacial thickness from the approach curve was 0.7 nm,

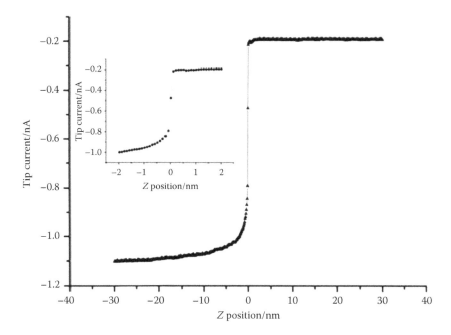

FIGURE 2.14 SICM approach curve of a nanopipette approached through a water/nitrobenzene interface. The nitrobenzene interface contains 1 mM TPAsTPB as the supporting electrolyte while the water phase contains 10 mM LiCl as the supporting electrolyte. The inset shows the middle portion of the approach curve, from −2 to 2 nm in the Z position. (Reproduced from Ji, T. et al., Probing the structure of a water/nitrobenzene interface by scanning ion conductance microscopy. *Chemical Science* **2011**, *2*, 1523–1529. Reproduced by Permission of The Royal Society of Chemistry.)

which was in agreement with other reports. The detection of a continuously fluctuating ion current at the interface was indicative of a mixed solvent layer that consisted of ions penetrating the interfacial region, which indicated the ITIES may not be molecularly sharp. Two important conclusions from this study include the determination of an interfacial thickness less than 1 nm and the suggestion that ions can penetrate into the interface. Although ITIES is typically examined with SECM, SICM has experimental advantages such as enhanced probe control, smaller probe dimensions, and less perturbation of the interface.

2.7.5 LIVING CELLS

Scanning ion conductance microscopy has been widely applied for the study of living cells due to the noncontact nature of the technique and ability to image under physiological conditions. A similar, widely used electrophysiological technique is the patch clamp, which uses a pipette to form a gigaohm seal against a cell to measure ion conductance.[94] The patch clamp is known to have high temporal resolution for measurement of membrane electrical activity of various excitable cells such as cardiomyocytes and neurons. When SICM and the patch clamp are combined to form the "smart" patch clamp system, both topography and single ion channel measurements can be recorded.[95–103] In contrast to conventional patch clamp systems coupled with light microscopy, where the relative position of the pipette and the cell surface are not well defined and hard to control, high-resolution scanning patch clamp enables a pipette to be patched at a specific position on the cell surface by first imaging the surface topography with SICM. In the first smart patch clamp report,[95] the spatial distribution of ATP-regulated K$^+$ ion channels in the cell plasma membrane of cardiac myocytes was mapped, as shown in Figure 2.15. Determination of ion channel distribution is important for elucidation of regional specialization within a cell for localized function. Ion channel clusters were found to be spaced 2–6 μm apart, while ion channels within the clusters were spaced 0.2–1 μm.[95] High-resolution scanning patch clamp techniques are successful for the localization and characterization of single ion channels, which is otherwise difficult.

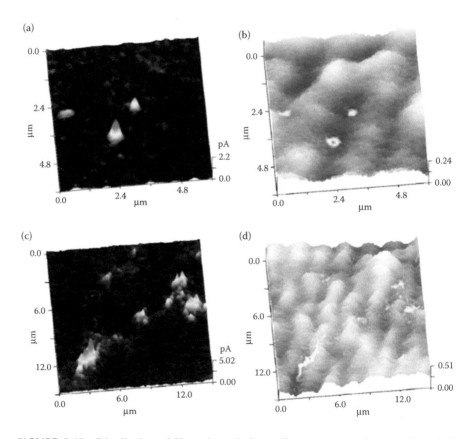

FIGURE 2.15 Distribution of K$_{ATP}$ channels in cardiomyocyte sarcolemma. (a) and (c) Maps of K$^+$ current and (b) and (d) corresponding topography superimposed on current map of a quiescent cardiac myocyte. (Reprinted from Korchev, Y. E. et al., Functional localization of single active ion channels on the surface of a living cell. *Nature Cell Biology* **2000**, 2, 616–619. Copyright 2008 with permission from Elsevier.)

A brief summary of SICM applications are listed in Table 2.2. Substrate, signal, and probe type for each reported SICM study through 2012 are listed.

2.8   COMPARISON OF IMAGING FORMATS

Scanning probe microscopy techniques commonly used for electrochemical measurements include SICM, SECM, and SECM-SICM. Each technique has distinct advantages and disadvantages. Important considerations are type(s) of signal detected, feedback mechanism, imaging conditions, relative ease of probe fabrication and resolution, many of which are interrelated. In the following section, SICM, SECM, and SECM-SICM will be considered relative to one another.

The most fundamental difference between SICM, SECM, and SECM-SICM is the type of signal detected in each technique. Scanning ion conductance microscopy can only detect ion conductance over a sample and is inherently "chemically blind." Some studies have shown that redox properties of[7] or through[18] a substrate can be correlated with ion conductance measurements but electrochemical imaging is still impossible for SICM in the original configuration. Other modifications have been made to enable potentiometric imaging.[17] In contrast, SECM is a powerful technique for faradaic measurements, but poor probe control inhibits topographic imaging resolution. These detection limitations of both instruments are overcome when SECM and SICM are combined into a hybrid technique. With SECM-SICM, electrochemical, ion current, and topographic images can be simultaneously recorded with similar resolution to the stand-alone techniques.[21]

Feedback mechanism is an important factor in SPM studies because probe control is required for sensitive measurements. The ion current feedback mechanism used in SICM enables precise control and positioning of the probe relative to the sample. Noncontact imaging is also a result of the robust feedback system. Typical applications of SECM employ less reliable feedback mechanisms, and the probe is often maintained at a constant height above the sample.[59] This condition can result in probe crash when the sample topography contains dynamic or tall structural features, or poor overall resolution. Feedback mechanisms utilized in SICM have been an important motivator for the hybridization of SICM with other techniques. For instance, when ion current feedback is combined with SECM in SECM-SICM, electrochemical imaging resolution is significantly enhanced. Ion current-based probe control has enabled high-resolution imaging of enzyme patterns for which SECM had previously been incapable.[19]

A practical limitation for SECM and SECM-SICM is probe fabrication. SICM pipettes can be produced in a quick, easy, and inexpensive manner. Multiple pipettes with highly controlled parameters can be prepared with high throughput. SECM probes, however, can be difficult to fabricate, especially with small dimensions. Typically, advanced instrumentation and time-consuming protocols are required. An even greater technical challenge can be the dual SECM-SICM probe. Integration of a small electrode onto an SICM pipette can be tedious, although simple methods such as the DBCNP protocol have been developed.[21] Additionally, care must be taken to avoid cross talk between the electronics of both components.

TABLE 2.2
SICM Applications

Application	Pipette Type	Signal Measured	Reference
AFM calibration grid	Single barrel	Ion current	55
Aluminum on glass	Hybrid probe	Ion current	72
Beads on glass	Hybrid probe	Ion current	44
Carbon nanotubes	Double barrel	Ion current; electron transfer	81
Compact disc	Single barrel	Ion current	46
Conducting polymer	Single barrel	Ion current	7
Cu/Sb on glass	Single barrel	Ion current	8
Deposition	Single barrel	N/A	32–34
Deposition	Double barrel	N/A	21–22
Enzyme spot	Single barrel	Ion current; electron transfer	19
Au electrode	Hybrid probe	Ion current; electron transfer	20
Au film on carbon	Single barrel	Ion current	43
Au film on glass	Hybrid probe	Ion current; electron transfer	20
Au film on glass	Single barrel	Ion current	57
Graphene	Double barrel	Ion current; electron transfer	81,86
HOPG	Double barrel	Ion current; electron transfer	83
Interdigitated electrode array	Single barrel	Ion current	19
ITIES	Single barrel	Ion transfer	9
Li$^+$ ion battery	Single barrel	Ion current	8
Living cells	Double barrel	Ion current; electron transfer	21
Living cells	Single barrel	Ion transfer	6, 19, 40, 42–43, 46, 48–55, 62–67, 75, 95–128
Living cells	Single barrel	Ion current; electron transfer	19
Living cells	Double barrel	Ion current, potential	17
Living cells	Hybrid probe	Ion current	68, 118
Polymer membrane	Single barrel	Ion current	4, 11–14, 36, 45–46
Polymer membrane	Hybrid probe	Electron transfer	18–19
Polymer membrane	Double barrel	Ion current, potential	
Pore suspending membrane	Double barrel	Electron transfer	129
Pt electrode	Double barrel	Ion current, electron transfer	84–85
Resolution theory	Single barrel	Ion current	38, 63
Sample manipulation	Single barrel	Ion current	78
SNOM	Single barrel	N/A	44, 67–74
Trapping	Single barrel	N/A	30–31, 130
UV diffraction grating	Single barrel	Ion current	55
Black lipid membrane	Single barrel	Ion current	131

Resolution among the three methods is dependent upon many factors. Generally, SICM is an inherently high-resolution technique because of small probe dimensions and robust feedback. High-resolution chemical imaging can be achieved with SECM as well, especially as UME dimensions decrease. Lack of D_{ps} control, though, can inhibit the quality of electrochemical images produced. The resolution of SECM-SICM is also complex. Studies have indicated topography and ion current resolution of the hybrid method is similar to SICM when the same sample is imaged with both techniques. However, many factors dictate resolution for the SECM portion of the probe. If the UME cannot be fabricated with small dimensions, such as when the electrode size is limited by the nanopipette opening in the SBDP technique, SECM resolution may be poor. With the DBCNP fabrication strategy, DBCNPs with radii as small as 10 nm have been produced.[21] Overall, SICM has practical advantages over SECM and SECM-SICM such as probe control and ease of pipette fabrication but other factors must be considered for each technique.

2.9 CONCLUSIONS

While SICM and SECM experiments were initially carried out on home-built systems, a number of commercial vendors for both techniques now exist. The first company to commercially manufacture SICM instruments was Ionscope (London, England). Ionscope's SICM series can be equipped with non-modulated, modulated, hopping mode, fluorescence, and electrophysiological capabilities. Additional features include a temperature and CO_2 controlled chamber for prolonged cellular imaging, easy integration with an inverted optical microscope, and FPGA incorporation for high performance.

In 2009, long-time AFM manufacturers Park Systems (Suwon, South Korea) developed an SICM instrument geared specifically toward live cell imaging. This system integrates SICM, noncontact AFM, and optical microscopy into one instrument. The system also allows for electrophysiological patch-clamp measurements and can function in approach retract scan mode, which is similar to hopping mode. A specialized environmental chamber for temperature, pH, and humidity control is available, which can help to keep cells viable for longer than 20 h.

Scanning electrochemical microscopy hardware is amenable to adaptation for SICM, with popular models produced by CH Instruments (Austin, TX). This microscope has a wide range of capabilities as well as features relevant to the electrochemist, such as a function generator and bipotentiostat. SECM experiments can be performed in constant height, constant current, potentiometric, and impedance modes. A number of SECM and multi-SPM instruments are also available through various vendors, such as HEKA (Lambrecht/Pfalz Germany). Besides these SECM-specific instruments, Uniscan Instruments (Buxton, United Kingdom) and Princeton Applied Research (Oakridge, TN) both produce electrochemical workstations that can be operated with SECM, scanning vibrating electrode technique, scanning kelvin probe, localized electrochemical impedance spectroscopy, scanning droplet system, and optical surface profiling.

To summarize the major concepts discussed in this chapter, the use of nanopipettes as the scanning probe in SICM enables high precision, resolution, and amenability to hybridization with other techniques. Probe fabrication is quick with high throughput and bare quartz or borosilicate pipettes are modified easily with other materials. Nanopipettes have special fluidic properties, which can be used for substrate characterization as well as dielectrophoretic trapping.

Scanning ion conductance microscopy, like other SPM techniques, has complex resolution considerations. Resolution is dependent upon probe–surface interaction, as well as probe geometry and probe–surface distance. Image convolutions and artifacts occur as a result of interactions between ion current distribution of the pipette tip and the sample surface. Various models have been proposed to elucidate this behavior.

Scanning ion conductance microscopy has a wide range of applications, the outlook for this technique is favorable, as continued probe development, and hybridization with other methods brings SICM closer to single molecule detection. Limitations for SICM will be realized as a finite limit for probe radii is reached, however other probe fabrication methods may overcome this obstacle. SICM is still relatively new, and exciting studies that make use of this technique in unique ways are reported continually. Continued feedback mode improvement has enabled high-resolution imaging of complicated surface structures. Methods that use multiple approach curves over a sample have been developed for improved biological imaging. Other methods which allow for unprecedentedly fast image acquisition have been created, as well. Additional microscopy methods have been paired with SICM to create dual functional techniques. Enhanced probe control, as well as high resolution of SICM makes hybridization of SICM with other techniques beneficial. Further, limitations of SICM, such as insensitivity to faradaic current, are overcome through hybridization with SECM.

In the past two decades, SICM has grown from a low-resolution technique with a micron-sized, eyedropper probe to a routinely used tool capable of multi-surface interrogation. The simple two-electrode setup and relatively trivial probe fabrication process have encouraged unique configurations and applications for SICM to be developed. Once only capable of micron resolution, the limits of SICM are pushed continually, as probe size decreases and elegant solutions to experimental obstacles are developed. Recent commercialization of SICM will further encourage technological advance. As the capability of SICM continues to grow with hybridizations to existing instruments, techniques such as SECM-SICM will foreseeably be commercialized in the future.

ACKNOWLEDGMENT

The authors thank Baker group members for thoughtful comments and discussion regarding the manuscript.

REFERENCES

1. Binnig, G.; Rohrer, H.; Gerber, C.; Weibel, E., Tunneling through a controllable vacuum gap. *Applied Physics Letters* **1982**, *40*, 178–180.
2. Bard, A. J.; Fan, F. R. F.; Kwak, J.; Lev, O., Scanning electrochemical microscopy. Introduction and principles. *Analytical Chemistry* **1989**, *61*, 132–138.

3. Engstrom, R. C.; Weber, M.; Wunder, D. J.; Burgess, R.; Winquist, S., Measurements within the diffusion layer using a microelectrode probe. *Analytical Chemistry* **1986**, *58*, 844–848.

4. Hansma, P.; Drake, B.; Marti, O.; Gould, S.; Prater, C., The scanning ion-conductance microscope. *Science* **1989**, *243*, 641–643.

5. Chen, C.-C.; Zhou, Y.; Baker, L. A., Scanning ion conductance microscopy. *Annual Review of Analytical Chemistry* **2012**, *5*, 207–228.

6. Happel, P.; Thatenhorst, D.; Dietzel, I. D., Scanning ion conductance microscopy for studying biological samples. *Sensors* **2012**, *12*, 14983–15008.

7. Laslau, C.; Williams, D. E.; Wright, B. E.; Travas-Sejdic, J., Measuring the ionic flux of an electrochemically actuated conducting polymer using modified scanning ion conductance microscopy. *Journal of the American Chemical Society* **2011**, *133*, 5748–5751.

8. Lipson, A. L.; Ginder, R. S.; Hersam, M. C., Nanoscale *In situ* characterization of Li-ion battery electrochemistry via scanning ion conductance microscopy. *Advanced Materials* **2011**, *23*, 5613–5617.

9. Ji, T.; Liang, Z.; Zhu, X.; Wang, L.; Liu, S.; Shao, Y., Probing the structure of a water/nitrobenzene interface by scanning ion conductance microscopy. *Chemical Science* **2011**, *2*, 1523–1529.

10. Zhou, L.; Zhou, Y.; Baker, L. A., Measuring ions with scanning ion conductance microscopy. *Electrochemical Society Interface* **2014**, *23*, 47–53.

11. Chen, C.-C.; Zhou, Y.; Baker, L. A., Single-nanopore investigations with ion conductance microscopy. *ACS Nano* **2011**, *5*, 8404–8411.

12. Zhou, Y.; Chen, C.-C.; Baker, L. A., Heterogeneity of multiple-pore membranes investigated with ion conductance microscopy. *Analytical Chemistry* **2012**, *84*, 3003–3009.

13. Chen, C.-C.; Baker, L. A., Effects of pipette modulation and imaging distances on ion currents measured with scanning ion conductance microscopy (SICM). *Analyst* **2011**, *136*, 90–97.

14. Chen, C.-C.; Derylo, M. A.; Baker, L. A., Measurement of ion currents through porous membranes with scanning ion conductance microscopy. *Analytical Chemistry* **2009**, *81*, 4742–4751.

15. Zhou, Y.; Chen, C.-C.; Weber, A. E.; Zhou, L.; Baker, L. A., Potentiometric-scanning ion conductance microscopy. *Langmuir* **2014**, *30*, 5669–5675.

16. Morris, C. A.; Chen, C.-C.; Ito, T.; Baker, L. A., Local pH Measurement with scanning ion conductance microscopy. *Journal of the Electrochemical Society* **2013**, *160*, H430–H435.

17. Chen, C.-C.; Zhou, Y.; Morris, C. A.; Hou, J.; Baker, L. A., Scanning ion conductance microscopy measurement of paracellular channel conductance in tight junctions. *Analytical Chemistry* **2013**, *85*, 3621–3628.

18. Morris, C. A.; Chen, C.-C.; Baker, L. A., Transport of redox probes through single pores measured by scanning electrochemical-scanning ion conductance microscopy (SECM-SICM). *Analyst* **2012**, *137*, 2933–2938.

19. Takahashi, Y.; Shevchuk, A. I.; Novak, P.; Murakami, Y.; Shiku, H.; Korchev, Y. E.; Matsue, T., Simultaneous noncontact topography and electrochemical imaging by SECM/SICM featuring ion current feedback regulation. *Journal of the American Chemical Society* **2010**, *132*, 10118–10126.

20. Comstock, D. J.; Elam, J. W.; Pellin, M. J.; Hersam, M. C., Integrated ultramicroelectrode-nanopipet probe for concurrent scanning electrochemical microscopy and scanning ion conductance microscopy. *Analytical Chemistry* **2010**, *82*, 1270–1276.

21. Takahashi, Y.; Shevchuk, A. I.; Novak, P.; Zhang, Y.; Ebejer, N.; Macpherson, J. V.; Unwin, P. R.; Pollard, A. J.; Roy, D.; Clifford, C. A., Multifunctional nanoprobes for nanoscale chemical imaging and localized chemical delivery at surfaces and interfaces. *Angewandte Chemie International Edition* **2011**, *50*, 9638–9642.

22. Rodolfa, K. T.; Bruckbauer, A.; Zhou, D.; Korchev, Y. E.; Klenerman, D., Two-component graded deposition of biomolecules with a double-barreled nanopipette. *Angewandte Chemie International Edition* **2005**, *44*, 6854–6859.

23. Wei, C.; Bard, A. J.; Feldberg, S. W., Current rectification at quartz nanopipet electrodes. *Analytical Chemistry* **1997**, *69*, 4627–4633.

24. Siwy, Z.; Apel, P.; Dobrev, D.; Neumann, R.; Spohr, R.; Trautmann, C.; Voss, K., Ion transport through asymmetric nanopores prepared by ion track etching. *Nuclear Instruments and Methods in Physics Research Section B: Beam Interactions with Materials and Atoms* **2003**, *208*, 143–148.

25. Siwy, Z.; Gu, Y.; Spohr, H. A.; Baur, D.; Wolf-Reber, A.; Spohr, R.; Apel, P.; Korchev, Y. E., Rectification and voltage gating of ion currents in a nanofabricated pore. *Europhysics Letters* **2002**, *60*, 349–355.

26. Siwy, Z. S., Ion-current rectification in nanopores and nanotubes with broken symmetry. *Advanced Functional Materials* **2006**, *16*, 735–746.

27. White, H. S.; Bund, A., Ion current rectification at nanopores in glass membranes. *Langmuir* **2008**, *24*, 2212–2218.

28. Woermann, D., Analysis of non-ohmic electrical current-voltage characteristic of membranes carrying a single track-etched conical pore. *Nuclear Instruments & Methods in Physics Research, Section B: Beam Interactions with Materials and Atoms* **2002**, *194*, 458–462.

29. Woermann, D., Electrochemical transport properties of a cone-shaped nanopore: High and low electrical conductivity states depending on the sign of an applied electrical potential difference. *Physical Chemistry Chemical Physics* **2003**, *5*, 1853–1858.

30. Clarke, R. W.; White, S. S.; Zhou, D.; Ying, L.; Klenerman, D., Trapping of proteins under physiological conditions in a nanopipette. *Angewandte Chemie* **2005**, *117*, 3813–3816.

31. Ying, L.; White, S. S.; Bruckbauer, A.; Meadows, L.; Korchev, Y. E.; Klenerman, D., Frequency and voltage dependence of the dielectrophoretic trapping of short lengths of DNA and dCTP in a nanopipette. *Biophysical Journal* **2004**, *86*, 1018–1027.

32. Bruckbauer, A.; Ying, L.; Rothery, A. M.; Zhou, D.; Shevchuk, A. I.; Abell, C.; Korchev, Y. E.; Klenerman, D., Writing with DNA and protein using a nanopipet for controlled delivery. *Journal of the American Chemical Society* **2002**, *124*, 8810–8811.

33. Ying, L.; Bruckbauer, A.; Rothery, A. M.; Korchev, Y. E.; Klenerman, D., Programmable delivery of DNA through a nanopipet. *Analytical Chemistry* **2002**, *74*, 1380–1385.

34. Bruckbauer, A.; James, P.; Zhou, D.; Yoon, J. W.; Excell, D.; Korchev, Y.; Jones, R.; Klenerman, D., Nanopipette delivery of individual molecules to cellular compartments for single-molecule fluorescence tracking. *Biophysical Journal* **2007**, *93*, 3120–3131.

35. Piper, J. D.; Li, C.; Lo, C.-J.; Berry, R.; Korchev, Y.; Ying, L.; Klenerman, D., Characterization and application of controllable local chemical changes produced by reagent delivery from a nanopipet. *Journal of the American Chemical Society* **2008**, *130*, 10386–10393.

36. Nitz, H.; Kamp, J.; Fuchs, H., A combined scanning ion-conductance and shear-force microscope. *Probe Microscopy* **1998**, *1*, 187–200.

37. Schraml, S. *Setup and Application of a Scanning Ion Conductance Microscope.* Vienna University of Technology, Vienna **2003**.

38. Edwards, M. A.; Williams, C. G.; Whitworth, A. L.; Unwin, P. R., Scanning ion conductance microscopy: A model for experimentally realistic conditions and image interpretation. *Analytical Chemistry* **2009**, *81*, 4482–4492.

39. Cornut, R.; Lefrou, C., A unified new analytical approximation for negative feedback currents with a microdisk SECM tip. *Journal of Electroanalytical Chemistry* **2007**, *608* (1), 59–66.

40. Korchev, Y. E.; Bashford, C. L.; Milovanovic, M.; Vodyanoy, I.; Lab, M. J., Scanning ion conductance microscopy of living cells. *Biophysical Journal* **1997**, *73*, 653–658.
41. Wipf, D. O.; Bard, A. J., Scanning electrochemical microscopy. 15. Improvements in imaging via tip-position modulation and lock-in detection. *Analytical Chemistry* **1992**, *64*, 1362–1367.
42. Shevchuk, A. I.; Gorelik, J.; Harding, S. E.; Lab, M. J.; Klenerman, D.; Korchev, Y. E., Simultaneous measurement of Ca^{2+} and cellular dynamics: Combined scanning ion conductance and optical microscopy to study contracting cardiac myocytes. *Biophysical Journal* **2001**, *81*, 1759–1764.
43. Li, C.; Johnson, N.; Ostanin, V.; Shevchuk, A.; Ying, L.; Korchev, Y.; Klenerman, D., High resolution imaging using scanning ion conductance microscopy with improved distance feedback control. *Progress in Natural Science* **2008**, *18*, 671–677.
44. Mannelquist, A.; Iwamoto, H.; Szabo, G.; Shao, Z., Near-field optical microscopy with a vibrating probe in aqueous solution. *Applied Physics Letters* **2001**, *78*, 2076–2078.
45. Proksch, R.; Lal, R.; Hansma, P. K.; Morse, D.; Stucky, G., Imaging the internal and external pore structure of membranes in fluid: TappingMode scanning ion conductance microscopy. *Biophysical Journal* **1996**, *71*, 2155–2157.
46. Böcker, M.; Anczykowski, B.; Wegener, J.; Schäffer, T. E., Scanning ion conductance microscopy with distance-modulated shear force control. *Nanotechnology* **2007**, *18*, 145505–145506.
47. Borgwarth, K.; Ebling, D. G.; Heinze, J., Scanning electrochemical microscopy: A new scanning mode based on convective effects. *Berichte der Bunsengesellschaft für physikalische Chemie* **1994**, *98*, 1317–1321.
48. Novak, P.; Li, C.; Shevchuk, A. I.; Stepanyan, R.; Caldwell, M.; Hughes, S.; Smart, T. G.; Gorelik, J.; Ostanin, V. P.; Lab, M. J., Nanoscale live-cell imaging using hopping probe ion conductance microscopy. *Nature Methods* **2009**, *6*, 279–281.
49. Takahashi, Y.; Murakami, Y.; Nagamine, K.; Shiku, H.; Aoyagi, S.; Yasukawa, T.; Kanzaki, M.; Matsue, T., Topographic imaging of convoluted surface of live cells by scanning ion conductance microscopy in a standing approach mode. *Physical Chemistry Chemical Physics* **2010**, *12*, 10012–10017.
50. Mann, S.; Hoffmann, G.; Hengstenberg, A.; Schuhmann, W.; Dietzel, I., Pulse-mode scanning ion conductance microscopy—A method to investigate cultured hippocampal cells. *Journal of Neuroscience Methods* **2002**, *116*, 113–117.
51. Happel, P.; Dietzel, I. D., Backstep scanning ion conductance microscopy as a tool for long term investigation of single living cells. *Journal of Nanobiotechnology* **2009**, *7*, 7–16.
52. Happel, P.; Hoffmann, G.; Mann, S.; Dietzel, I., Monitoring cell movements and volume changes with pulse-mode scanning ion conductance microscopy. *Journal of Microscopy* **2003**, *212*, 144–151.
53. Mann, S.; Meyer, J.; Dietzel, I., Integration of a scanning ion conductance microscope into phase contrast optics and its application to the quantification of morphological parameters of selected cells. *Journal of Microscopy* **2006**, *224*, 152–157.
54. Mann, S. A.; Versmold, B.; Marx, R.; Stahlhofen, S.; Dietzel, I. D.; Heumann, R.; Berger, R., Corticosteroids reverse cytokine-induced block of survival and differentiation of oligodendrocyte progenitor cells from rats. *Journal of Neuroinflammation* **2008**, *5*, 39.
55. Zhukov, A.; Richards, O.; Ostanin, V.; Korchev, Y.; Klenerman, D., A hybrid scanning mode for fast scanning ion conductance microscopy (SICM) imaging. *Ultramicroscopy* **2012**, *121*, 1–7.
56. Novak, P.; Shevchuk, A.; Ruenraroengsak, P.; Miragoli, M.; Thorley, A. J.; Klenerman, D.; Lab, M. J.; Tetley, T. D.; Gorelik, J.; Korchev, Y. E., Imaging single nanoparticle interactions with human lung cells using fast ion conductance microscopy. *Nano Letters* **2014**, *14*, 1202–1207.

57. McKelvey, K.; Perry, D.; Byers, J. C.; Colburn, A. W.; Unwin, P. R., Bias modulated scanning ion conductance microscopy. *Analytical Chemistry* **2014**, *86*, 3639–3646.

58. Schwarz, U. D.; Hölscher, H.; Wiesendanger, R., Atomic resolution in scanning force microscopy: Concepts, requirements, contrast mechanisms, and image interpretation. *Physical Review B* **2000**, *62*, 13089.

59. Amemiya, S.; Bard, A. J.; Fan, F.-R. F.; Mirkin, M. V.; Unwin, P. R., Scanning electrochemical microscopy. *Annual Review of Analytical Chemistry* **2008**, *1*, 95–131.

60. Rayleigh, L., XXXI. Investigations in optics, with special reference to the spectroscope. *The London, Edinburgh, and Dublin Philosophical Magazine and Journal of Science* **1879**, *8*, 261–274.

61. Wong, C.; West, P.; Olson, K.; Mecartney, M.; Starostina, N., Tip dilation and AFM capabilities in the characterization of nanoparticles. *Jouranl of Metals* **2007**, *59*, 12–16.

62. Shevchuk, A. I.; Frolenkov, G. I.; Sánchez, D.; James, P. S.; Freedman, N.; Lab, M. J.; Jones, R.; Klenerman, D.; Korchev, Y. E., Imaging proteins in membranes of living cells by high-resolution scanning ion conductance microscopy. *Angewandte Chemie* **2006**, *118*, 2270–2274.

63. Rheinlaender, J.; Schäffer, T. E., Image formation, resolution, and height measurement in scanning ion conductance microscopy. *Journal of Applied Physics* **2009**, *105*, 094905.

64. Gorelik, J.; Shevchuk, A.; Ramalho, M.; Elliott, M.; Lei, C.; Higgins, C.; Klenerman, D.; Krauzewicz, N.; Korchev, Y., Scanning surface confocal microscopy for simultaneous topographical and fluorescence imaging: Application to single virus-like particle entry into a cell. *Proceedings of the National Academy of Sciences* **2002**, *99*, 16018–16023.

65. Shevchuk, A. I.; Hobson, P.; Lab, M. J.; Klenerman, D.; Krauzewicz, N.; Korchev, Y. E., Imaging single virus particles on the surface of cell membranes by high-resolution scanning surface confocal microscopy. *Biophysical Journal* **2008**, *94*, 4089–4094.

66. Shevchuk, A. I.; Hobson, P.; Klenerman, D.; Krauzewicz, N.; Korchev, Y. E., Endocytic pathways: Combined scanning ion conductance and surface confocal microscopy study. *Pflügers Archiv-European Journal of Physiology* **2008**, *456*, 227–235.

67. Kemp, S. J.; Thorley, A. J.; Gorelik, J.; Seckl, M. J.; O'Hare, M. J.; Arcaro, A.; Korchev, Y.; Goldstraw, P.; Tetley, T. D., Immortalization of human alveolar epithelial cells to investigate nanoparticle uptake. *American Journal of Respiratory Cell and Molecular Biology* **2008**, *39*, 591–597.

68. Shevchuk, A. I.; Novak, P.; Taylor, M.; Diakonov, I. A.; Ziyadeh-Isleem, A.; Bitoun, M.; Guicheney, P.; Gorelik, J.; Merrifield, C. J.; Klenerman, D., An alternative mechanism of clathrin-coated pit closure revealed by ion conductance microscopy. *The Journal of Cell Biology* **2012**, *197*, 499–508.

69. Betzig, E.; Isaacson, M.; Barshatzky, H.; Lin, K.; Lewis, A., Near field scanning optical microscopy, Proc. *SPIE* **1988**, *987*, 91–99.

70. Müller, A.-D.; Müller, F.; Hietschold, M., Electrochemical pattern formation in a scanning near-field optical microscope. *Applied Physics A: Materials Science & Processing* **1998**, *66*, S453–S456.

71. Korchev, Y. E.; Raval, M.; Lab, M. J.; Gorelik, J.; Edwards, C. R.; Rayment, T.; Klenerman, D., Hybrid scanning ion conductance and scanning near-field optical microscopy for the study of living cells. *Biophysical Journal* **2000**, *78*, 2675–2679.

72. Mannelquist, A.; Iwamoto, H.; Szabo, G.; Shao, Z., Near field optical microscopy in aqueous solution: Implementation and characterization of a vibrating probe. *Journal of Microscopy* **2002**, *205*, 53–60.

73. Rothery, A.; Gorelik, J.; Bruckbauer, A.; Yu, W.; Korchev, Y.; Klenerman, D., A novel light source for SICM–SNOM of living cells. *Journal of Microscopy* **2003**, *209*, 94–101.

74. Bruckbauer, A.; Ying, L.; Rothery, A. M.; Korchev, Y. E.; Klenerman, D., Characterization of a novel light source for simultaneous optical and scanning ion conductance microscopy. *Analytical Chemistry* **2002**, *74*, 2612–2616.

75. Pellegrino, M.; Pellegrini, M.; Orsini, P.; Tognoni, E.; Ascoli, C.; Baschieri, P.; Dinelli, F., Measuring the elastic properties of living cells through the analysis of current–displacement curves in scanning ion conductance microscopy. *Pflügers Archiv-European Journal of Physiology* **2012**, *464*, 307–316.

76. Sánchez, D.; Johnson, N.; Li, C.; Novak, P.; Rheinlaender, J.; Zhang, Y.; Anand, U.; Anand, P.; Gorelik, J.; Frolenkov, G. I., Noncontact measurement of the local mechanical properties of living cells using pressure applied via a pipette. *Biophysical Journal* **2008**, *95*, 3017–3027.

77. Sánchez, D.; Anand, U.; Gorelik, J.; Benham, C. D.; Bountra, C.; Lab, M.; Klenerman, D.; Birch, R.; Anand, P.; Korchev, Y., Localized and non-contact mechanical stimulation of dorsal root ganglion sensory neurons using scanning ion conductance microscopy. *Journal of Neuroscience Methods* **2007**, *159*, 26–34.

78. Pellegrino, M.; Orsini, P.; Pellegrini, M.; Baschieri, P.; Dinelli, F.; Petracchi, D.; Tognoni, E.; Ascoli, C., Weak hydrostatic forces in far-scanning ion conductance microscopy used to guide neuronal growth cones. *Neuroscience Research* **2011**, *69*, 234–240.

79. Güell, A. G.; Ebejer, N.; Snowden, M. E.; McKelvey, K.; Macpherson, J. V.; Unwin, P. R., Quantitative nanoscale visualization of heterogeneous electron transfer rates in 2D carbon nanotube networks. *Proceedings of the National Academy of Sciences* **2012**, *109*, 11487–11492.

80. Güell, A. G.; Ebejer, N.; Snowden, M. E.; Macpherson, J. V.; Unwin, P. R., Structural correlations in heterogeneous electron transfer at monolayer and multilayer graphene electrodes. *Journal of the American Chemical Society* **2012**, *134*, 7258–7261.

81. Lai, S.; Patel, A. N.; McKelvey, K.; Unwin, P. R., Definitive evidence for fast electron transfer at pristine basal plane graphite from high-resolution electrochemical imaging. *Angewandte Chemie* **2012**, *124*, 5501–5504.

82. Ebejer, N.; Schnippering, M.; Colburn, A. W.; Edwards, M. A.; Unwin, P. R., Localized high resolution electrochemistry and multifunctional imaging: Scanning electrochemical cell microscopy. *Analytical Chemistry* **2010**, *82*, 9141–9145.

83. Patten, H. V.; Lai, S. C.; Macpherson, J. V.; Unwin, P. R., Active sites for outer-sphere, inner-sphere, and complex multistage electrochemical reactions at polycrystalline boron-doped diamond electrodes (pBDD) revealed with scanning electrochemical cell microscopy (SECCM). *Analytical Chemistry* **2012**, *84*, 5427–5432.

84. Aaronson, B. D.; Chen, C.-H.; Li, H.; Koper, M. T.; Lai, S. C.; Unwin, P. R., Pseudo-single-crystal electrochemistry on polycrystalline electrodes: Visualizing activity at grains and grain boundaries on platinum for the Fe2+ /Fe3+ redox reaction. *Journal of the American Chemical Society* **2013**, *135*, 3873–3880.

85. Chen, C.-H.; Meadows, K. E.; Cuharuc, A.; Lai, S. C.; Unwin, P. R., High resolution mapping of oxygen reduction reaction kinetics at polycrystalline platinum electrodes. *Physical Chemistry Chemical Physics* **2014**, *16*, 18545–18552.

86. Kirkman, P. M.; Guell, A. G.; Cuharuc, A. S.; Unwin, P. R., Spatial and temporal control of the diazonium modification of sp2 carbon surfaces. *Journal of the American Chemical Society* **2013**, *136*, 36–39.

87. Zhang, G.; Kirkman, P. M.; Patel, A. N.; Cuharuc, A. S.; McKelvey, K.; Unwin, P. R., Molecular functionalization of graphite surfaces: Basal plane vs step edge electrochemical activity. *Journal of the American Chemical Society* **2014**, *136*, 11444–11451.

88. Aaronson, B. D.; Lai, S. C.; Unwin, P. R., Spatially resolved electrochemistry in ionic liquids: Surface structure effects on triiodide reduction at platinum electrodes. *Langmuir* **2014**, *30*, 1915–1919.

89. Ebejer, N.; Güell, A. G.; Lai, S. C.; McKelvey, K.; Snowden, M. E.; Unwin, P. R., Scanning electrochemical cell microscopy: A versatile technique for nanoscale electrochemistry and functional imaging. *Annual Review of Analytical Chemistry* **2013**, *6*, 329–351.

90. Nadappuram, B. P.; McKelvey, K.; Al Botros, R.; Colburn, A. W.; Unwin, P. R., Fabrication and characterization of dual function nanoscale pH-scanning ion conductance microscopy (SICM) probes for high resolution pH mapping. *Analytical Chemistry* **2013**, *85*, 8070–8074.

91. Gavach, C.; Seta, P.; D'epenoux, B., The double layer and ion adsorption at the interface between two non miscible solutions: Part I. Interfacial tension measurements for the water-nitrobenzene tetraalkylammonium bromide systems. *Journal of Electroanalytical Chemistry and Interfacial Electrochemistry* **1977**, *83*, 225–235.

92. Girault, H.; Schiffrin, D., Thermodynamic surface excess of water and ionic solvation at the interface between immiscible liquids. *Journal of Electroanalytical Chemistry and Interfacial Electrochemistry* **1983**, *150*, 43–49.

93. Wei, C.; Bard, A. J.; Mirkin, M. V., Scanning electrochemical microscopy. 31. Application of SECM to the study of charge transfer processes at the liquid/liquid interface. *The Journal of Physical Chemistry* **1995**, *99*, 16033–16042.

94. Sakmann, B.; Neher, E., Patch clamp techniques for studying ionic channels in excitable membranes. *Annual Review of Physiology* **1984**, *46*, 455–472.

95. Korchev, Y. E.; Negulyaev, Y. A.; Edwards, C. R.; Vodyanoy, I.; Lab, M. J., Functional localization of single active ion channels on the surface of a living cell. *Nature Cell Biology* **2000**, *2*, 616–619.

96. James, A. F.; Sabirov, R. Z.; Okada, Y., Clustering of protein kinase A-dependent CFTR chloride channels in the sarcolemma of guinea-pig ventricular myocytes. *Biochemical and Biophysical Research Communications* **2010**, *391*, 841–845.

97. Gu, Y.; Gorelik, J.; Spohr, H. A.; Shevchuk, A.; Harding, S. E.; Vodyanoy, I.; Klenerman, D.; Korchev, Y. E., High-resolution scanning patch-clamp: New insights into cell function. *The FASEB Journal* **2002**, *16*, 748–750.

98. Gorelik, J.; Zhang, Y.; Sánchez, D.; Shevchuk, A.; Frolenkov, G.; Klenerman, D.; Edwards, C.; Korchev, Y., Aldosterone acts via an ATP autocrine/paracrine system: The Edelman ATP hypothesis revisited. *Proceedings of the National Academy of Sciences of the United States of America* **2005**, *102*, 15000–15005.

99. Dutta, A. K.; Korchev, Y. E.; Shevchuk, A. I.; Hayashi, S.; Okada, Y.; Sabirov, R. Z., Spatial distribution of maxi-anion channel on cardiomyocytes detected by smart-patch technique. *Biophysical Journal* **2008**, *94*, 1646–1655.

100. Yang, X.; Liu, X.; Lu, H.; Zhang, X.; Ma, L.; Gao, R.; Zhang, Y., Real-time investigation of acute toxicity of ZnO nanoparticles on human lung epithelia with hopping probe ion conductance microscopy. *Chemical Research in Toxicology* **2012**, *25*, 297–304.

101. Miragoli, M.; Moshkov, A.; Novak, P.; Shevchuk, A.; Nikolaev, V. O.; El-Hamamsy, I.; Potter, C. M.; Wright, P.; Kadir, S. S. A.; Lyon, A. R., Scanning ion conductance microscopy: A convergent high-resolution technology for multi-parametric analysis of living cardiovascular cells. *Journal of the Royal Society Interface* **2011**, *8*, 913–925.

102. Gorelik, J.; Gu, Y.; Spohr, H. A.; Shevchuk, A. I.; Lab, M. J.; Harding, S. E.; Edwards, C. R.; Whitaker, M.; Moss, G. W.; Benton, D. C., Ion channels in small cells and subcellular structures can be studied with a smart patch-clamp system. *Biophysical Journal* **2002**, *83*, 3296–3303.

103. Duclohier, H., Neuronal sodium channels in ventricular heart cells are localized near T-tubules openings. *Biochemical and Biophysical Research Communications* **2005**, *334*, 1135–1140.

104. Chen, X.; Zhu, H.; Liu, X.; Lu, H.; Li, Y.; Wang, J.; Liu, H.; Zhang, J.; Ma, Q.; Zhang, Y., Characterization of two mammalian cortical collecting duct cell lines with hopping probe ion conductance microscopy. *The Journal of Membrane Biology* **2013**, *246*, 7–11.

105. Gorelik, J.; Harding, S. E.; Shevchuk, A. I.; Koralage, D.; Lab, M.; de Swiet, M.; Korchev, Y.; Williamson, C., Taurocholate induces changes in rat cardiomyocyte contraction and calcium dynamics. *Clinical Science* **2002**, *103*, 191–200.

106. Gorelik, J.; Shevchuk, A.; Swiet, M.; Lab, M.; Korchev, Y.; Williamson, C., Comparison of the arrhythmogenic effects of tauro- and glycoconjugates of cholic acid in an *in vitro* study of rat cardiomyocytes. *BJOG: An International Journal of Obstetrics & Gynaecology* **2004**, *111*, 867–870.

107. Gorelik, J.; Shevchuk, A.; Diakonov, I.; Swiet, M.; Lab, M.; Korchev, Y.; Williamson, C., Dexamethasone and ursodeoxycholic acid protect against the arrhythmogenic effect of taurocholate in an *in vitro* study of rat cardiomyocytes. *BJOG: An International Journal of Obstetrics & Gynaecology* **2003**, *110*, 467–474.

108. Gorelik, J.; Shevchuk, A. I.; Frolenkov, G. I.; Diakonov, I. A.; Kros, C. J.; Richardson, G. P.; Vodyanoy, I.; Edwards, C. R.; Klenerman, D.; Korchev, Y. E., Dynamic assembly of surface structures in living cells. *Proceedings of the National Academy of Sciences* **2003**, *100*, 5819–5822.

109. Gorelik, J.; Vodyanoy, I.; Shevchuk, A. I.; Diakonov, I. A.; Lab, M. J.; Korchev, Y. E., Esmolol is antiarrhythmic in doxorubicin-induced arrhythmia in cultured cardiomyocytes–determination by novel rapid cardiomyocyte assay. *FEBS Letters* **2003**, *548*, 74–78.

110. Gorelik, J.; Yang, L. Q.; Zhang, Y.; Korchev, Y.; Harding, S. E., A novel Z-groove index characterizing myocardial surface structure. *Cardiovascular Research* **2006**, *72*, 422–429.

111. Happel, P.; Möller, K.; Schwering, N. K.; Dietzel, I. D., Migrating oligodendrocyte progenitor cells swell prior to soma dislocation. *Scientific Reports* **2013**, *3*, 1806.

112. Ibrahim, M.; Kukadia, P.; Siedlecka, U.; Cartledge, J. E.; Navaratnarajah, M.; Tokar, S.; Van Doorn, C.; Tsang, V. T.; Gorelik, J.; Yacoub, M. H., Cardiomyocyte Ca^{2+} handling and structure is regulated by degree and duration of mechanical load variation. *Journal of Cellular and Molecular Medicine* **2012**, *16*, 2910–2918.

113. Klenerman, D.; Korchev, Y. E.; Davis, S. J., Imaging and characterisation of the surface of live cells. *Current Opinion in Chemical Biology* **2011**, *15*, 696–703.

114. Korchev, Y.; Milovanovic, M.; Bashford, C.; Bennett, D.; Sviderskaya, E.; Vodyanoy, I.; Lab, M., Specialized scanning ion-conductance microscope for imaging of living cells. *Journal of Microscopy* **1997**, *188*, 17–23.

115. Korchev, Y. E.; Gorelik, J.; Lab, M. J.; Sviderskaya, E. V.; Johnston, C. L.; Coombes, C. R.; Vodyanoy, I.; Edwards, C. R., Cell volume measurement using scanning ion conductance microscopy. *Biophysical Journal* **2000**, *78*, 451–457.

116. Liu, B.-C.; Song, X.; Lu, X.-Y.; Fang, C. Z.; Wei, S.-P.; Alli, A. A.; Eaton, D. C.; Shen, B.-Z.; Li, X.-Q.; Ma, H.-P., Lovastatin attenuates effects of cyclosporine A on tight junctions and apoptosis in cultured cortical collecting duct principal cells. *American Journal of Physiology-Renal Physiology* **2013**, *305*, F304–F313.

117. Lorin, C.; Gueffier, M.; Bois, P.; Faivre, J.-F.; Cognard, C.; Sebille, S., Ultrastructural and functional alterations of EC coupling elements in mdx cardiomyocytes: An analysis from membrane surface to depth. *Cell Biochemistry and Biophysics* **2013**, *66*, 723–736.

118. Lyon, A. R.; MacLeod, K. T.; Zhang, Y.; Garcia, E.; Kanda, G. K.; Korchev, Y. E.; Harding, S. E.; Gorelik, J., Loss of T-tubules and other changes to surface topography in ventricular myocytes from failing human and rat heart. *Proceedings of the National Academy of Sciences* **2009**, *106*, 6854–6859.

119. Lyon, A. R.; Nikolaev, V. O.; Miragoli, M.; Sikkel, M. B.; Paur, H.; Benard, L.; Hulot, J.-S.; Kohlbrenner, E.; Hajjar, R. J.; Peters, N. S., Plasticity of surface structures and β2-adrenergic receptor localization in failing ventricular cardiomyocytes during recovery from heart failure. *Circulation: Heart Failure* **2012**, *5*, 357–365.

120. Miragoli, M.; Novak, P.; Ruenraroengsak, P.; Shevchuk, A. I.; Korchev, Y. E.; Lab, M. J.; Tetley, T. D.; Gorelik, J., Functional interaction between charged nanoparticles and cardiac tissue: A new paradigm for cardiac arrhythmia? *Nanomedicine* **2013**, *8*, 725–737.

121. Mizutani, Y.; Choi, M.-H.; Cho, S.-J.; Okajima, T., Nanoscale fluctuations on epithelial cell surfaces investigated by scanning ion conductance microscopy. *Applied Physics Letters* **2013**, *102*, 173703–173708.

122. Nikolaev, V. O.; Moshkov, A.; Lyon, A. R.; Miragoli, M.; Novak, P.; Paur, H.; Lohse, M. J.; Korchev, Y. E.; Harding, S. E.; Gorelik, J., β2-Adrenergic receptor redistribution in heart failure changes cAMP compartmentation. *Science* **2010**, *327*, 1653–1657.

123. Potter, C. M.; Schobesberger, S.; Lundberg, M. H.; Weinberg, P. D.; Mitchell, J. A.; Gorelik, J., Shape and compliance of endothelial cells after shear stress *in vitro* or from different aortic regions: Scanning ion conductance microscopy study. *PloS One* **2012**, *7*, e31228.

124. Rheinlaender, J.; Geisse, N. A.; Proksch, R.; Scha¨ffer, T. E., Comparison of scanning ion conductance microscopy with atomic force microscopy for cell imaging. *Langmuir* **2010**, *27*, 697–704.

125. Shin, W.; Gillis, K. D., Measurement of changes in membrane surface morphology associated with exocytosis using scanning ion conductance microscopy. *Biophysical Journal* **2006**, *91*, L63–L65.

126. Zhang, S.; Cho, S.-J.; Busuttil, K.; Wang, C.; Besenbacher, F.; Dong, M., Scanning ion conductance microscopy studies of amyloid fibrils at nanoscale. *Nanoscale* **2012**, *4*, 3105–3110.

127. Zhang, Y.; Sanchez, D.; Gorelik, J.; Klenerman, D.; Edwards, C.; Korchev, Y., Basolateral P2X4-like receptors regulate the extracellular ATP-stimulated epithelial Na+ channel activity in renal epithelia. *American Journal of Physiology-Renal Physiology* **2007**, *292*, F1734–F1740.

128. Zhang, Y.; Gorelik, J.; Sanchez, D.; Shevchuk, A.; Lab, M.; Vodyanoy, I.; Klernerman, D.; Edwards, C.; Korchev, Y., Scanning ion conductance microscopy reveals how a functional renal epithelial monolayer maintains its integrity. *Kidney International* **2005**, *68*, 1071–1077.

129. Böcker, M.; Muschter, S.; Schmitt, E. K.; Steinem, C.; Schäffer, T. E., Imaging and patterning of pore-suspending membranes with scanning ion conductance microscopy. *Langmuir* **2009**, *25*, 3022–3028.

130. Jönsson, P.; McColl, J.; Clarke, R. W.; Ostanin, V. P.; Jönsson, B.; Klenerman, D., Hydrodynamic trapping of molecules in lipid bilayers. *Proceedings of the National Academy of Sciences* **2012**, *109*, 10328–10333.

131. Höfer, I.; Steinem, C., A membrane fusion assay based on pore-spanning lipid bilayers. *Soft Matter* **2011**, *7*, 1644–1647.

3 Electrode Surface Modification Using Diazonium Salts

Avni Berisha, Mohamed M. Chehimi,
Jean Pinson, and Fetah I. Podvorica

CONTENTS

3.1   AN OVERVIEW OF THE PROCESS

Diazonium salts have been known for a very long time (Peter Griess in 1858), and they have been widely used for the preparation of azo dyes. A novel reaction of these compounds has been evidenced in 1992: the modification of surfaces by aryl groups [1]; it constitutes the topic of this chapter and can be summarized by Figure 3.1. In this very simplified scheme, a surface is used as a cathode in an electrochemical cell containing a solution of an aromatic diazonium salt in acetonitrile (ACN) or acidic water. A one electron reduction at potentials close to 0 V/SCE (saturated calomel electrode) leads to the homolytic dediazonation of the diazonium cation. The result-ing aryl radical (Ar·) binds to the surface through a covalent bond providing a modi-fied surface [2–8]. Diazonium salts are obtained from amines in the presence of a nitrosating reagent (NO$^+$ obtained in different ways, Table 3.1) through a very simple reaction; they can be isolated as crystalline solids but they can also be prepared *in situ* and used without any need for isolation [9].

Since many aromatic amines are commercially available, a large number of diazonium salts can be synthetized and grafted on surfaces. If the diazonium is substituted by a reactive R group, the surface can be further modified. Using this method, surfaces can be obtained that are modified with simple organic groups as well as proteins and polymers (Section 3.8). Numerous applications from sensors to composites can be engineered from such surfaces. In addition to using electrochemi-cal reduction, photochemical activation, or chemical reduction implying electron transfer in order to achieve dediazonation and subsequent grafting, other methods can also be employed [15] such as ultrasonication or heating. The reaction can be performed in organic solvents, mainly ACN, but also in aqueous medium which is an attractive feature in view of further applications. A number of surfaces can be modified by metals (Au, Pt, Ni, Fe, etc.), semiconductors (Si, Ge, etc.), and insu-lating substrates (SiO$_2$, TiN, TaN, organic polymers) using non-electrochemical dediazonation methods. After the discovery of this reaction, there was some doubt about the existence of a real covalent bond between the surface and the aryl groups. However, the demonstration of the robustness of the assembly as a function of dif-ferent parameters (temperature, solvents, mechanical stress, etc.) and theoretical and

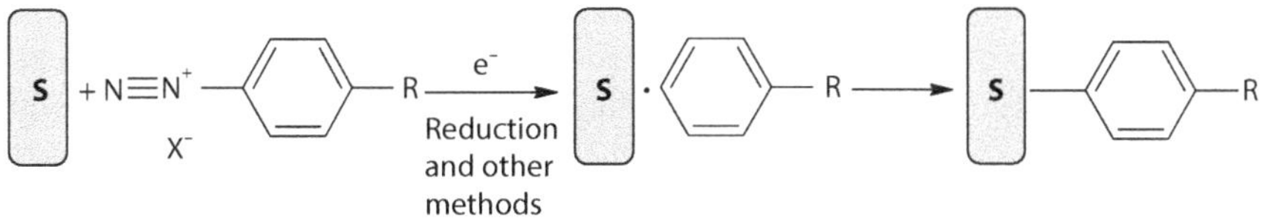

FIGURE 3.1   A schematic presentation of the grafting of surfaces by diazonium chemistry.

TABLE 3.1
Some Methods for the Preparation of Diazonium Salts

The classical method in aqueous solution:

From amines in aqueous 25% HBF_4 in the presence of 1.1 mole equivalent of $NaNO_2$, used without isolation [9] or filtered, rinsed with ether, dried under vacuum, and stored at $-20°C$.

If the amine is substituted by an acid group, addition of an acid can be omitted [10]:

With aminobenzoic, -benzenesulfonic, and -benzylphosphonic acids in water in the presence of $NaNO_2$, without isolation

NO_2^- *can be produced in situ by reduction of NO_3^- on a Cu or carbon electrode* [11]:

In the presence of amines in acidic medium, the diazonium salt is formed and directly grafted on the Cu electrode

From triazenes in the presence of an electrogenerated acid [12]:

The electrochemical oxidation of N,N'-diphenylhydrazine produces the protons necessary for the cleavage of the triazene, a diazonium is formed and directly attached to the glassy carbon (GC) surface

In aprotic medium [13]: From amines, in CH_3CN, in the presence of 1.1 equivalent of $NOBF_4$ at $-20°C$, precipitated in the presence of Et_2O, filtered, dried under vacuum, and stored at $-20°C$

Without solvent [14]: From amines, in the presence of 1.1 equivalent of isoamylnitrite at $60°$ in the presence of the surface

---

spectroscopic evidence firmly established the existence of this bond. For simplicity, Figure 3.1 presents a single layer of aromatic groups. In most cases (i.e., if no special conditions are used), multilayers are obtained with a thickness from a few nanometer to micrometer. This reaction has led to a large number of academic papers that recently focused on the surface modification of nanomaterials; it has also reached the industrial stage and new developments are ongoing.

The surface chemistry of diazonium salts is only the most recent of important organic reactions of these species [15]. These reactions are summarized in Figure 3.2. Most of them are related to the homolytic dediazonation and therefore bear some connection with the grafting under examination.

Reaction [R1] is the surface modification, the subject of this chapter. With a few exceptions discussed below, it is assigned to the attack of *aryl radicals* on the surface. Reaction [R2] is the Balz–Schiemann reaction [16] that transforms a diazonium tetrafluoroborate into a fluoroaromatic upon heating. Hence tetrafluoroborates are preferred to other counter anions; they suppress the risk of explosions of the diazonium salt. Reaction [R3] is the Sandmeyer reaction [17–18] where, in the presence of the reducing reagents CuBr and Cl, the diazonium salt provides an *aryl radical* that abstracts Br and Cl atoms to give ArBr and Cl. Reaction [R4] is the Meerwein reaction [19] where the intermediate is also an *aryl radical* that attacks an alkene; the *aryl radical* can also be produced by electrochemical reduction of an aryl halide [20]. Reaction [R5] is the Gomberg Bachman reaction [21–22] which occurs when an *aryl radical* reacts with an aromatic. It is responsible for the formation of by-products such as biaryls during the grafting and electrografting of surfaces. These by-products precipitate on the surface which is why a careful rinsing is mandatory before examining modified surfaces by spectroscopy. For reaction [R6],

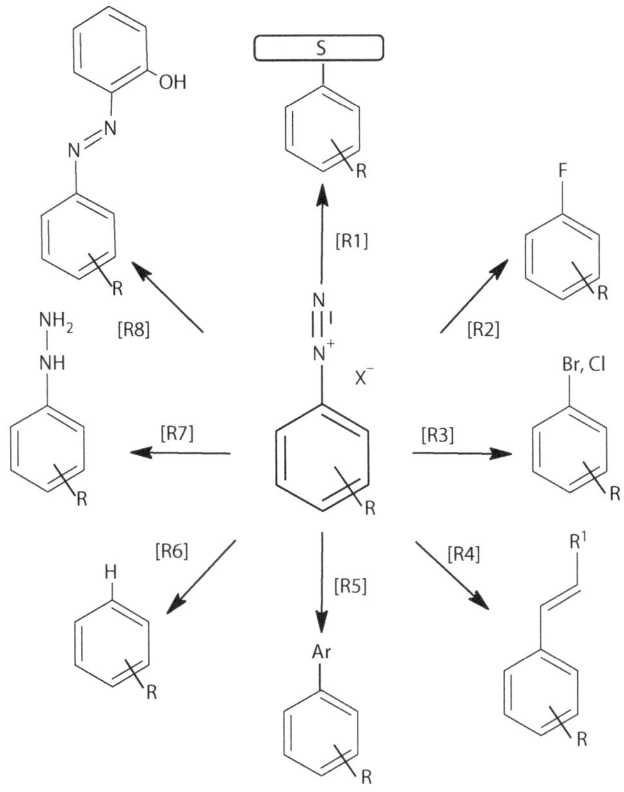

FIGURE 3.2 Some reactions of diazonium salts (see text).

if a diazonium cation is reduced to a radical, far from any surface and with no other reagent present in the solution, it will abstract a hydrogen from the solvent (e.g., ACN) to give ArH and the ·CH$_2$CN radical that also reacts with the surface [23–24]. In aprotic medium, a diazonium cation is reduced by one electron to the *aryl radical* responsible for reaction [R1]. However, in aqueous medium, on a second three elec-tron wave, a hydrazine is formed following reaction [R7]. Reaction [R8] corresponds to the heterolytic cleavage of diazonium salts, which occurs upon heating in water. This azo C-coupling reaction with nucleophiles such as phenols is the basis of dye chemistry.

Depending on the medium, diazonium salts can be transformed into different species that can also be used for electrografting. These reactions are summarized in Figure 3.3. The different species that can be obtained in aqueous or alcoholic solu-tions give rise to complex equilibria. Therefore, before electrografting or chemical grafting, one should determine which species is present in the solution used and its lifetime. Diazonium salts are relatively stable in aqueous acidic solution (pH < 4), where dediazonation proceeds through the rate-limiting formation of a highly unstable aryl cation Ar$^+$ that reacts immediately with nucleophiles present in the

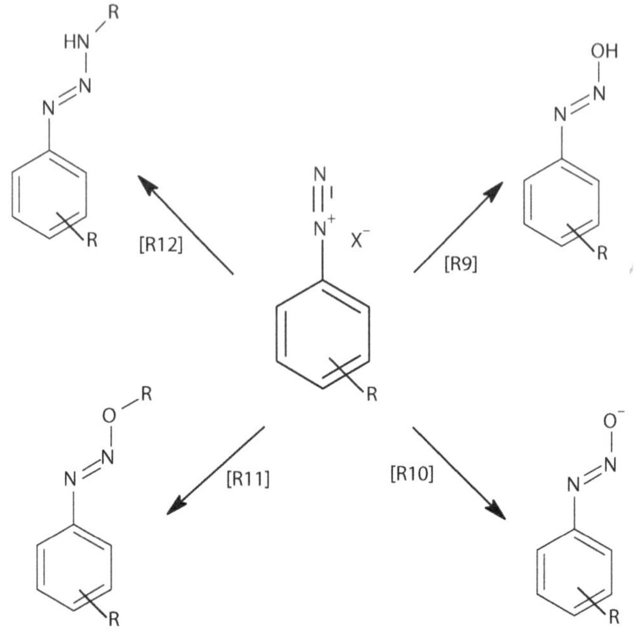

FIGURE 3.3   Different derivatives of diazonium salts of interest in electrografting.

solution. If the pH increases above pK 5–6, diazohydroxides are formed [R9]; at pH > 9, diazoates are obtained [R10] [25]. These species can also be reduced and grafted. In alcohol, diazoethers arc obtained [R11] [26–28]. Diazonium salts react with secondary amines [R12] to give triazenes [29] that are stable in basic media but give rise to *aryl radicals* under acidic solutions; their stability is orthogonal to that of diazonium salts.

The next sections will develop and elaborate on the points raised above. The final section will present other electrografting similar to that of diazonium salts. Where possible, diazonium salts will be designed by **D** followed by the formula of the substituent; for example, **DNO$_2$** for 4-nitrophenylbenzenediazonium. The film–surface–film assembly will be sketched as *Au*–**NO$_2$** for a gold surface modified by nitrophenyl groups and the films will be designed as polyphenylene films, even if we show below that their structure is more complex. Single or double wall carbon nanotubes will be referred to as *S* or *DWCNT*–**NO$_2$** and graphene as *Gr*–**NO$_2$**; nanoparticles (nP) as, for example, Au(nP) and *Au(nP)*–**NO$_2$**; nanorods by Au(nRod) and *Au(nRod)*–**NO$_2$**.

3.2   THE ELECTROCHEMISTRY OF DIAZONIUM SALTS

The electrochemistry of diazonium salts was first examined by polarography [30–31], where it was possible to establish the presence of two waves in aqueous

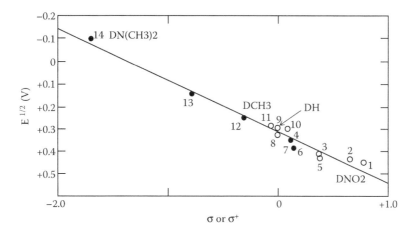

FIGURE 3.4 Relation of half-wave potentials/SCE for the reduction of substituted benzenediazonium salts in sulfolane to the Hammett substituent constant. (Adapted from Elofson, R. M. and F. F. Gadallah. *J. Org. Chem.* 34, 1969: 854–857.)

acidic medium: a first wave at ~+ 0.2 V/SCE for **DCH₃** corresponding to one electron transfer and leading to a radical, and a second 3e⁻ wave at more negative potentials resulting in an overall 4e⁻ reduction to phenylhydrazine. In addition, an adsorption phenomenon was noted on the first wave and the half wave potentials were found to be linearly related to the Hammett constant of the substituents (Figure 3.4). Very interestingly, in agreement with previous results [32], it was found that coulometry on a mercury pool provided nearly exclusively phenylmercuric chloride and diphenylmercury; their formation was assigned to the reaction of phenyl radicals with mercury. This reaction can be viewed as "electrografting on mercury." The formation of radicals was also observed by esr after electrochemical reduction of **DH** on a mercury electrode and fouling of a Pt electrode [33]. Later on, the electrochemical behavior of aryldiazonium salts was investigated at metallic electrodes such as platinum, gold, and mercury [34]. From their study, the authors concluded that the radicals generated during the potential scan led to the blocking of the electrode surface; however, the nature of the resulting blocking layer was not investigated. Therefore, at the beginning of the 1970s, the main characteristics of the reduction of diazonium salts that permit electrografting had already been established.

The electrochemistry of diazonium salts can therefore be discussed in terms of two reactions (Figure 3.5).

$$ArN_2^+ + 1e^- \longrightarrow Ar^{\cdot} \quad [R13]$$

$$Ar^{\cdot} + 1e^- \longrightarrow Ar^- \quad [R14]$$

FIGURE 3.5 Formation and reduction of an aryl radical.

3.2.1 Polarography

Figure 3.6 shows differential pulse polarograms of the first wave at pH values of 3.43, 6.04, and 8.14 (where the diazonium cations, the diazohydroxide, and the diazoate are, respectively, the dominant species) [25].

3.2.2 Cyclic Voltammetry

Cyclic voltammetry permits a different view of the process. Figure 3.7 shows cyclic voltammograms of **DNO₂** and **DCl** in ACN and a protic ionic liquid [35–36]. These voltammograms show the typical characteristics of diazonium salts in aprotic or acidic protic media: (i) a broad irreversible wave, (ii) the reduction potential is close to 0 V/SCE, (iii) the height corresponds approximately to the transfer of 1e⁻, and (iv) the wave disappears upon repetitive scanning due to the electrografting process itself. As the potential is cycled, the electrode is covered by an organic layer that prevents electron transfer. Additional features that appear on the voltammogram of **DNO₂** (Figure 3.7c) include a prewave and a small reversible wave on the return scan. On gold, the prewave has been assigned to the reduction of the diazonium salt on different crystallographic sites of the electrode; on carbon, a similar explanation could be possible (Figure 3.8b) [37]. The small reversible system is assigned to the hydroxylamino/nitroso system resulting from the partial reduction of the nitro group to hydroxylamine (4e⁻ + 4H⁺). In addition, the inset of Figure 3.7 shows a redox probe experiment which indicates that the reversible voltammogram of the $Fe(CN)_6^{3-/4-}$ couple is partly blocked by the organic layer attached to the surface of the electrode (redox probe experiments are presented in more detail in Section 3.6). On the second wave of diazonium salts, phenylhdrazine is obtained in protic medium [30,31]. In aprotic medium (Hg cathode, ACN, −1.0 V/SCE), an anion is obtained that reacts

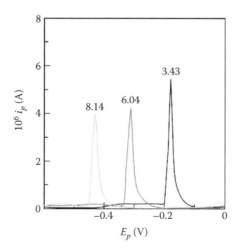

FIGURE 3.6 Polarograms of **DH** at different pH. (Adapted from Sienkiewicz, A., M. Szymula, J. Narkiewicz-Michalek, and C. Bravo-Díaz. *J. Phys. Org. Chem.* 27, 2014: 284–289.)

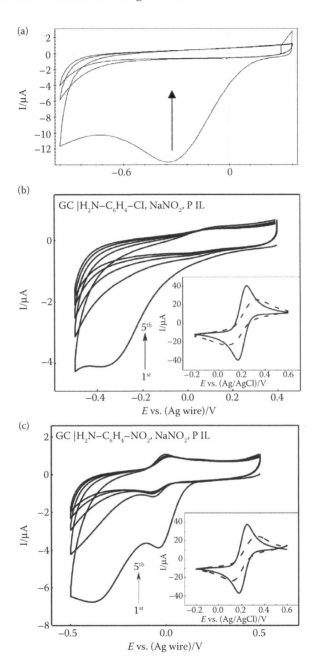

FIGURE 3.7 Repetitive cyclic voltammograms on GC of: (a) **DNO₂** prepared *ex situ*, in ACN; (b) **DCl**; and (c) **DNO₂**, both in prepared protic ionic liquid (2-methoxypyridine + tri-fluoroacetic acid) *in situ*. Inset: cyclic voltammograms of bare (*GC*) and electrografted electrodes by using (b) **DCl** and (c) **DNO₂** (dashed) immersed into 5 mM Fe(CN)$_6^{3-/4-}$/0.1 M KCl (pH 5). (Adapted from Le Floch, F. et al. *Mol. Cryst. Liq. Cryst.* 486, 2008: 271[1313]–281[1323]; Shul, G. et al. *Electrochim. Acta* 106, 2013: 378–385.)

with the diazonium salt to give azobenzene derivatives [38] without any formation of organomercury compounds.

Figure 3.8a and b shows voltammograms of **DNO$_2$** on a monolayer of graphene [39] in ACN and on gold, respectively; grafting on the different crystalline facets can be observed on gold [37]. Once the electrodes have been electrografted, it is possible to observe the groups attached to the surface if they are electroactive. Figure 3.9 shows the reversible voltammogram of a **GC** electrode modified with a quinone group [40].

3.2.3 CHRONOAMPEROMETRY

The grafting process can also be detected using chronoamperometry. Figure 3.10 is typical of what is observed with diazonium salts. A very sharp decrease of the current is observed for 3,5-dimethylbenzenediazonium (3,5-**D(CH$_3$)$_2$**). It is very different

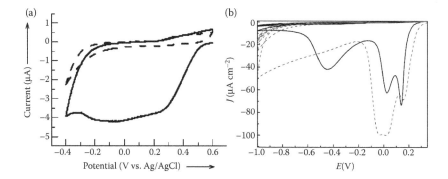

FIGURE 3.8 Cyclic voltammograms of (a) **DNO$_2$** on single layer of graphene in 0.1 M H$_2$SO$_4$. (b) **DSC$_6$F$_{13}$** (solid line) and of **DNO$_2$** +(dashed line) on **Au** substrates in ACN, reference Ag/AgClO$_4$. (Adapted from Gan, L., D. Zhang, and X. Guo. *Small* 8, 2012: 1326–1330; Benedetto, A. et al. *Electrochim. Acta* 53, 2008: 7117–7122, for (a) and (b), respectively.)

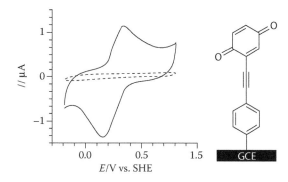

FIGURE 3.9 Cyclic voltammetry of **GC-quinone** into 0.1 M phosphate buffer pH 7.0 (solid line) and of the related electrochemically inactive hydroquinone (dashed line). (Adapted from Clausmeyer, J. *Chem. Phys. Chem.* 15, 2014: 151–156.)

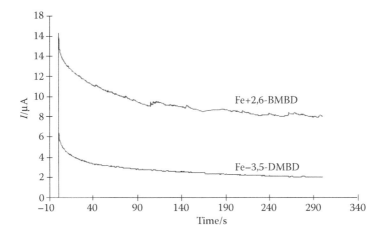

FIGURE 3.10 Chronoamperometry curve for an iron electrode immersed into a 10^{-2} M H_2SO_4 solution containing 3,5- or 2,6-**D(CH$_3$)$_2$**. Potential of the iron electrode: −0.7 V/SCE. (Adapted from Berisha, A. et al. *Electrochim. Acta* 56, 2011: 10762–10766.)

from the decrease observed for 2,6-dimethylbenzenediazonium (2,6-**D(CH$_3$)$_2$**) that does not lead to surface modification due to the steric hindrance of the two methyl groups [41].

Which is best for grafting diazonium salts: chonoamperometry or cyclic voltammetry? Cyclic voltammetry has been found to be more efficient during the grafting of thick nitrophenyl films. Potential sweeping was found to be essential to continuously desorb any physisorbed species that otherwise would clog the channels in the film and make it insulating [42].

3.2.4 SCANNING ELECTROCHEMICAL MICROSCOPY

Scanning ElectroChemical Microscopy (SECM) permits both the grafting and the characterization of the surface. For example, a surface was micropatterned by SECM through the process presented in Figure 3.11: (i) a nitrobenzene precursor is reduced at the Pt tip (25 µm) of the SECM to aniline at −0.4 V for 5 s, (ii) this aniline is readily transformed into benzenediazonium cation by $NaNO_2$ introduced in the solution, and (iii) this cation diffuses to the substrate polarized at −0.1 V/SCE, where it is reduced leading to the formation of a spot of derivatized **GC-R** [43]. This spot can then be imaged through an SECM line scan, recorded with $K_4Fe(CN)_6$ before (curve 1) and after (curve 2) the local derivatization. The current decrease is consistent with the formation of a 40-mm wide organic pattern that passivates the sample. Other examples of SECM examination of modified films are given in Section 3.6.

3.2.5 ELECTROCHEMICAL QUARTZ CRYSTAL MICROBALANCE

Electrochemical quartz crystal microbalance (EQCM) permits the observation of surface grafting through the decrease in frequency of (for example) the gold-coated

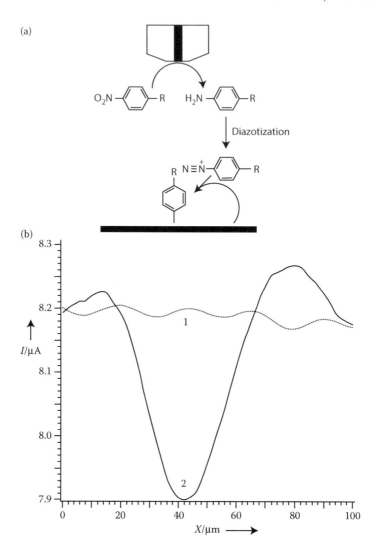

FIGURE 3.11 Schematic description of the patterning of a GC surface by SECM—(a) grafting of aryl moieties derived from diazonium salts. (b) SECM line scan with $K_4Fe(CN)_6$ (1) before and (2) after local modification of the sample. (Adapted from Cougnon, C. et al. *Ang. Chem. Int. Ed.* 48, 2009: 4006–4009.)

quartz that is related to the deposited mass. This mass can be transformed into surface concentration and, as in Figure 3.12, into monolayers equivalent [44]. However, this method is unable to discriminate between the mass of grafted and just deposited molecules, which makes the interpretation of the results more difficult. In Figure 3.12, the authors have used a very low concentration to limit the problem of deposited material (at very low concentrations, the relative yield of dimers and products resulting from bimolecular reactions should be lowered). The mass deposited increases with the concentration of **DNO$_2$**.

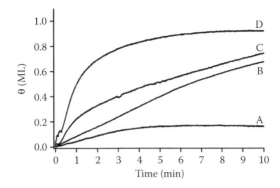

FIGURE 3.12   Surface coverage (expressed in monolayers equivalent) as a function of time up to 10 min for 5 µM (trace A), 10 µM (trace B), 50 µM (trace C), and 100 µM (trace D) **DNO$_2$** concentrations. (Adapted from Jayasundara, D. R. et al. *Langmuir* 27, 2011: 13029–13036.)

3.2.6   REDOX POTENTIAL OF THE ARYL RADICAL

The redox potential of the aryl radical is an important parameter in the electrochemical grafting. At more negative potentials, the radical will be reduced to an anion (Figure 3.5 [R14]). Several values have been estimated for this radical, either by experimentally or by density functional theory (DFT) calculations, and are listed in Table 3.2.

The first experimental $E°(C_6H_5\cdot/C_6H_5^-)$ value [46] was obtained by simulation of the voltammogram at low concentration ($c = 0.45$ mM) where, in this case, the reduction potential is the voltammetric peak potential. The second $E°$ value [46] is found by measuring the number of electrons transferred versus the reduction potential for a series of phenyldiazonium, iodonium, and sulfonium, where the reduction potential is located at the inflexion point of the sigmoidal curve. The three calculated values [47–49] were obtained by different DFT methods, taking in account the solvation. The values of $E°$ appear in the −0.8 to −1.0 V/SCE range in ACN with the exception of the value at −0.05 V/SCE measured by voltammetry that seems to be too positive (this may be due to some grafting during the voltammetry in

TABLE 3.2

Reduction Potentials of Aryl Radicals (V/SCE) in ACN, DMF, and THF

References	[45] Experimental	[46] Experimental	[47] Calculated	[48] Calculated	[49] Calculated
Reduction potentials $C_6H_5\cdot/C_6H_5^-$	ACN: −0.64	ACN: −0.95			
$E°(C_6H_5\cdot/C_6H_5^-)$	ACN: +0.05		ACN: −0.82	ACN: −0.95	ACN: −0.90
			DMF: −0.74		DMF: −0.81
					THF: −1.28

spite of the low concentration used). These values can also be compared with other experiments:

a. The oxidation potential of $C_6Br_5^-$ has been measured at +0.19 V/SCE, a quite positive potential even taking in account the effect of the bromines [49].
b. Diphenyl iodonium that is reduced at −1.0 V/SCE in ACN can be grafted on *GC* electrodes. This means that the phenyl radical is not entirely reduced to the anion at this potential.
c. The formation of carbanion-derived products (diarylazo) on mercury at −1.0 V/SCE indicates that the reduction of the *aryl radical* takes place at this potential [38].
d. At the potential where the radical is reduced, electrografting should become less efficient. However, to the best of our knowledge, there is only one report of a decreasing thickness or surface concentration as the potential is shifted to more and more negative values reaching and exceeding the redox potential of the radical. Figure 3.13 shows that the mass deposited on the EQCM electrode is smaller at increasingly more negative potentials, and the film is more homogeneous [50]. One possible interpretation of this last experiment (i.e., in contradiction with many other results; see Section 3.6) could be the reduction of the radical to the anion at potentials close to 0 V/SCE. However, it may also be due to the formation of a more blocking film with this particular diazonium salt. This point is discussed further in Section 3.6 in the context of the film growth. Additional experiments should therefore

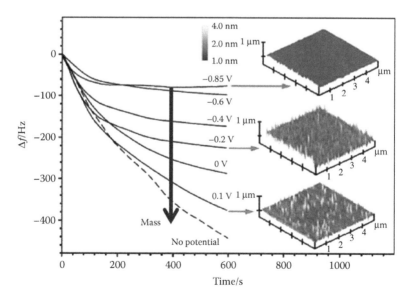

FIGURE 3.13   Mass deposited by EQCM in ACN on an Au electrode from **D-(CH$_2$)$_2$NH$_2$**, Δf, and atomic force microscopy (AFM) images of the surface (in ACN). (Adapted from Fau, M. et al. *Anal. Chem.* 83, 2011: 9281–9288.)

be performed to measure $E°(C_6H_5^./C_6H_5^-)$ under conditions where no grafting occurs for comparison with theoretical values, and also to observe less-efficient grafting at negative potentials.

3.3 NON-ELECTROCHEMICAL GRAFTING

In addition to electrochemical methods, grafting by diazonium salts can also be achieved by various other methods [51]. Although the exact grafting mechanisms of all the different procedures have not yet been investigated in detail, they can be sorted into two categories. The first include those for which an electron transfer is either demonstrated or the most probable path, while other mechanisms are likely involved in the second category. We first examine the methods involving an electron transfer that are by far the most numerous.

3.3.1 REDUCING SURFACES

The method closest to electrochemical reduction uses galvanic cells. It is possible to attach aryl groups by reduction of diazonium salts by contacting two metals, where the current flows from the less noble to the nobler metal (cathode). In this way, starting from **DNO₂**, it was possible to obtain quite thick polyphenylene-like films on gold in contact with an alligator clip; the process works better in ACN than in water [52]. With reducing metals as surfaces, grafting occurs spontaneously. For example, dipping a Fe, Ni, Cu, Zn coupon in a 10-mM solution of **DNO₂** in ACN is sufficient to obtain grafted films with thickness up to ~10 nm [53–55]. It has been shown that grafting is favored on reducing metals (by comparing Ni, Zn, Fe with $OCP^{ACN} = +0.30, -0.10, -0.50$, respectively) and with easily reduced diazonium salts. On copper, the spontaneous grafting of **DNO₂** was shown to be more efficient in ACN than in aqueous acidic solution [55]. A comparison of spontaneous grafting was also achieved between Cu° and oxidized copper (Cu_2O), where it was possible to detect by XPS the presence of Cu–aryl and Cu–O–aryl bonds [56].

Platinum can be reduced to negative oxidation states with oxidation potentials ranging from ~ −1 to ~ −2.5 V/SCE. This reduced Pt surface is spontaneously grafted through an electron transfer [57–58].

Spontaneous grafting is also observed on carbon black [59], but only with the most easily reducible diazonium salts, that is, **DNO₂**, **DCF₃**, or **DBr** ($E_p^{ACN} = +0.20, -0.34, -0.35$ V/SCE, respectively, where E_p^{ACN} refers to the peak potential measured by cyclic voltammetry. Spontaneous grafting is not observed with **DN(C₆H₅)₂**, **DNH(C₆H₅)**, **DNEt2** ($E_p^{ACN} = -0.60, -0.50$, and -0.56, respectively, on GC) [60]. This indicates an electron transfer process that is also supported by the dramatic change in potential observed in chronopotentiometry on a *GC* electrode upon addition of **DNO₂** (Figure 3.14).

After the initial growth of the polyphenylene film on the surface of carbon by a radical mechanism, a slower reaction permits a further growth of the film. It has been tentatively assigned to the attack of a spontaneously formed aryl cation on the first formed film [61] but it could also correspond to the chain mechanism described in Section 3.7. Spontaneous grafting on carbon was also investigated on amorphous

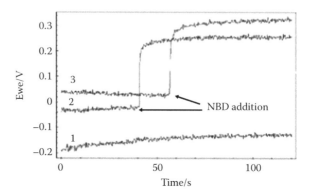

FIGURE 3.14 (1) Open-circuit potential (OCP) of a GC electrode in ACN. (2 and 3) A potential step is observed after addition of **DNO₂**. (Adapted from LeFloch, F., J.-P. Simonato, and G. Bidan. *Electrochim. Acta* 54, 2009: 3078–3085.)

carbon materials that differ in their sp^3-hydrogenated/sp^2 content. It was found that the surface concentration and initial rates decrease with the sp^3 content, that is, with the amount of hydrogenation. They are always lower with amorphous carbon than with *GC* [62].

Although carbon nanotubes can be modified by electrochemistry [63], most of the studies involve spontaneous derivatization. An electron transfer has been proposed in a diazonium/SWNT complex [64] but a detailed investigation of the mechanism in aqueous solution (i.e., in the presence of diazohydroxides) assigned the formation of aryl radicals (evidenced by trapping with TEMPO) to a spontaneous dediazonation [65].

Graphene can be spontaneously derivatized by diazonium salts [66–67]. The reaction is fast for a single layer of graphene but slower for a bilayer [68]. The grafting efficiency also depends on the substrate on which graphene is deposited [69]. Most papers describe, without a clear demonstration, this step as an electron transfer. But intercalation of alkali metal atoms in between the layers of graphite creates a very reducing reagent and addition of diazonium salts in the solutions permits at the same time to exfoliate graphite and modify the graphene sheets with aryl groups [70].

Spontaneous grafting with **DNO₂** is also possible on silicon. With different types of Si (mono or polycrystalline, highly n or p), the reaction occurs whatever the nature of the substrate, both in the dark and under illumination, with the best yield being obtained with highly doped n polysilicon under illumination [13,71–72]. In the same way, Ge can be spontaneously derivatized with diazonium salts [73–75].

3.3.2 REDUCTION BY CHEMICAL REAGENTS

Since diazonium salts are easily reduced, a number of weak reducing agents such as ascorbic acid [76], hypophosphorous acid [77–78], and Fe or Cu powder [79] can reduce the diazonium to an aryl radical that attaches to surfaces.

3.3.3 PHOTOCHEMISTRY

Diazonium salts can be reduced to radicals by irradiation of a charge transfer complex (diazonium cation/electron donor) under UV. With dimethoxybenzene as an electron donor, it is possible to initiate an electron transfer within the charge transfer complex, reduce the diazonium cation **DNO$_2$** to an aryl radical, and functionalize a gold surface with nitrophenyl groups [80–81]. Photografting is also possible in the presence of visible light; the mechanism is based on an electron transfer between an excited state of the sensitizer and the diazonium cation [82–85].

3.3.4 ULTRASONICATION

Indium tin oxide (ITO) has been modified by 4-nitrobenzenediazonium, in water, at different frequencies; the lowest one (20 kHz) was shown to be the most efficient. A mechanism based on the known formation of reducing radicals (H, R from adventitious impurities) has been proposed; these radicals should reduce the diazonium salt [86]. Therefore, this method can also be classified as involving electron transfer.

3.3.5 NEUTRAL OR BASIC pH

As pH increases, diazonium salts are progressively transformed to diazohydroxides and diazoates. These species undergo spontaneous dediazonation leading to an aryl radical, the key species of the grafting reaction [25–28]. The thermal dediazonation of **DBr** in aqueous acid solution is extremely slow ($t_{1/2} = 3000$ min at $T = 45°C$) but faster at pH 6 in 25% MeOH ($t_{1/2} = $ a few minutes) [87]; for **DNO$_2$** in 5% MeOH, $t_{1/2}$ decreases from 170 min at pH $= 0$ to a few minutes at pH $= 4$. This permits the spontaneous grafting of diazoates on gold [88]. However, diazoates are also easily reducible (Figure 3.6); for example, the diazoate from **DBr** is reduced at +0.07 V/SCE at pH 10, and therefore easily reduced by metals such Fe [88]. However, on silica nanoparticles, the spontaneous dediazonation permits the grafting of a polymerization initiator [89]. With diazoates, two parallel mechanisms are involved: a spontaneous decomposition and an electron transfer if the surface is reductive enough.

The following procedures, most probably, do not involve an electron transfer.

3.3.6 HEATING AND MICROWAVES

DC$_{10}$F$_{21}$ was prepared *in situ* by mixing the corresponding amine with neat isoamylnitrite directly on the surface of polymethylmethacrylate (PMMA); the surface was then heated to 60° for 1 h which resulted in a 2- to 4-layer *PMMA*-**C10F21** film [90]. A microwave-assisted surface modification of carbon nanotubes (CNTs) and nanohorns has also been described [91–93].

Therefore, there are many procedures available for the non-electrochemical grafting of diazonium salts, most of them likely involving an electron transfer. This versatility is one of the strong points of surface modification by diazonium salts.

3.4   THE SURFACE–ARYL BOND

The formation of a bond between the surface (metal, carbon, semiconductor), and the aryl group is a distinct feature of the process because it confers stability and the robustness to the assembly; it is the basis of many applications of the diazonium chemistry of surfaces. We now examine the different experiments that demonstrate the strong bonding of the organic layer through different bonds. It must be remembered that along with the grafting of aryl groups on the surface, the reduction of diazonium salts produces a number of side products such as dimers [21,76,94] that deposit on the surface; therefore, samples must be carefully rinsed before spectroscopic examination. For example, refluxing in ACN for 1 h an as-prepared Au–NO_2 sample removes 64% of the deposited material. In our experience, rinsing nanoparticles is particularly difficult due to their tendency to aggregate.

3.4.1   AMBIENT STABILITY

Organic films derived from diazonium salts are very stable; they remain unchanged for months when left on a bench [2,95]. After removal of the non-grafted material, Au–NO_2 samples resist refluxing for 1 h in ACN, ultrasonication for 30 min in ACN, and poly p-phenylene films grafted on gold (i.e., Au–$(C_6H_5)_n$) resist 3 h refluxing in toluene [96]. The stability of diazonium-derived films has been compared with that of thiols self assembled monolayers (SAMs) under different conditions: rigorous sonication, refluxing solvents, and chemical displacement by octadecanethiol. Under certain conditions, aryl films formed from the reduction of diazonium salts are more strongly bonded to gold surfaces compared to the thiol analogue [97,98]. The mechanical stability of the Si–NO_2 bond has been tested by AFM; even after scratching the multilayer, the signal of a very thin layer can be detected by XPS [99]. Si–Br and Si–NO_2 films resist 40% HF for 2 min and 10 M NH_4F for 1 min [100].

3.4.2   ELECTROCHEMICAL STABILITY

The electrochemical stability of diazonium-derived organic films is also an indication of the bonding between the surface and the film. This stability was tested with an electrode modified with diazonium-derived films ($DCOOH$, DNO_2, $DN(C_2H_5)_2$, and DBr) where the reversible voltammogram of the ferro/ferricyanide redox couple is blocked. This reversible system is restored if the film is removed. In this way, it was shown that very negative (–2 V/SCE) or positive (+1.8 V/SCE) potentials are necessary to remove the film. The same phenomenon could be observed by measuring the charge transfer resistance (Figure 3.15). Even then, XPS is able to detect grafted groups at the surface of the electrode [101]. By comparison, thiols SAMs are desorbed at much lower potentials (approx. –1 V/SCE for H–$(CH_2)_9$–SH [102]).

3.4.3   THERMAL STABILITY

The thermal stability of polyphenylene films has been examined on different surfaces. On highly oriented pyrolytic graphite (HOPG), it was shown by Auger spectroscopy,

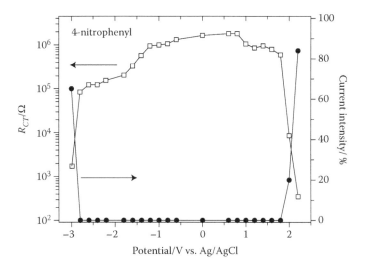

FIGURE 3.15   Variation of charge transfer resistance • and the current intensity □ as a function of the applied potentials on a *GC*-NO₂ electrode. (Adapted from D'Amours, M. and D. Bélanger. *J. Phys. Chem. B* 107, 2003: 4811–4817.)

under vacuum, that the O and N bands of **HOPG**–**NO₂** remained intact up to temperatures more than 400°C [2] and that they completely disappeared at temperatures more than 1000°C. Chlorobenzyl groups on GC: **GC**–**CH₂Cl** remain stable up to 400°C [103]. On carbon powder **Vulcan XC72R-NO₂**, in air, the main weight loss that occurs between 360°C and 500°C is assigned to the cleavage of the nitrophenyl group (as evidenced by TG-ms) just before the burn-off of the carbon particles [104]. Single wall carbon nanotubes derivatized with 4-chlorophenyl groups **SWCNT**–**Cl** lose 40% of their weight by heating to 500°C, with the main loss occurring above 400°C [105]. With **SWCNT-tBu**, the pristine **SWCNT** can be regenerated by heating to 300°C [106]. On stainless steel (SS), the **SS**–**NO₂** film is destroyed by heating up to 300–350°C [107]. With Al nanoparticles modified by diazonium chemistry with a photopolymerization initiator **Al**–**CH₂**–**S**–**C(=S)**–**N(Et)₂**, the coating is destroyed in the 200–400°C range [108]. Silica particles grafted in basic aqueous medium with the same photopolymerization initiator loses ~6% weight in the 300–400°C range [89].

Therefore, the thermal stability of these polyphenylene films is high with a decomposition temperature always higher than 200°C, whatever the substrate. By comparison, thiols that are used for surface modification desorb at lower temperature (e.g., HO–(CH₂)₁₈–SH desorbs at 150°C) [109].

3.4.4   X-Ray Photoelectron Spectroscopy

Direct observation of **Surface–aryl** bonds can be achieved in some cases such as **Metal-aryl**, **Metal–oxygen–aryl**, and **Metal–N = N–aryl**. For example, on iron **Fe–COOH**, a small C1s peak at 283.3 eV, at the foot of the main C1s peak, has been assigned to an iron carbide; that is, to the **Fe-aryl** bond [110]. As confirmation, it

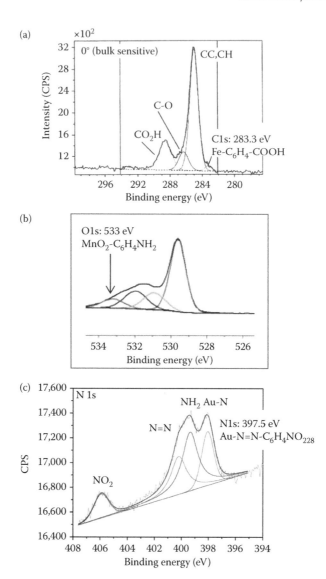

FIGURE 3.16 XPS evidence for (a) *Metal*–aryl, (b) *Metal*–O–aryl, and (c) *Metal*–N=N–aryl bonds during the grafting of diazonium salts. (Adapted from Boukerma, K. et al. *Langmuir* 19, 2003: 6333–6335; Bell, K. J. et al. *Chem. Comm.* 50, 2014: 13687–13690; Mesnage, A. et al. *Langmuir* 28, 2012: 11767–11778.)

was shown that this small peak decreased from 2.1% to 0.9% by tilting the x-ray beam toward the surface, indicating that the bond is located at the Fe-organic film interface. On oxides, *Surface*–**O**–**aryl** bonds have been observed on *MnO₂*–**NH₂** by the O1s signal at 533 eV [111]. More recently, bonding through *Au*–**N** = **N–C₆H₄NO₂** has been observed through a N1s peak at 397.5 eV assigned to an Au–N bond [112]. These results are presented in Figure 3.16 and tabulated in Table 3.3.

TABLE 3.3
Evidence of *Surface*–Aryl, *Surface*-O-Aryl, and *Surface*-N=N-Aryl Bonds by XPS

Modified Surfaces	Binding Energy (eV)	Comments References
Metal-aryl carbon bonds		
Grafting conditions[a]		
E: *Fe*–**COOH**	C1s: 283.3	[110]
E: *Fe*–**NO₂**	C1s: 283.8	[113]
S: *Ni*–**CH₃**	C1s: 283.5	[112]
E: *SS*–**CH₂–PO(OH)₂**	C1s: 282.9	SS: Stainless steel [114]
C: *Ag(np)*–**COOH**	Ag 3d5/2, 3d3/2: 366 and 372	Pristine Ag at 368 eV (3d5/2) and 374 eV (3d3/2) [115]
E: *Au*–**NH₂**	C1s: 284.1	[116]
E: *Au*–**COOH**	C1s: 283.5	[113]
E: *Au*–**Br**	C1s: Not detected	[117]
E: *SS*-**C(=O)C6H5**	C1s: ~283	[118]
Metal–N=N– bonds		
S: *Au*–**N=N–C₆H₄–NO₂**	N1s: 397.5 eV	[112,113]
S: *Ni*–**N=N–C₆H₄–OMe**	N1s: 397.7 eV	[112]
C: *V(np)*–**N=N–C₈H₁₇**	V2p: 514 N1s: 398.1 N1s: 396.6	[119]
E: *Au*–**N=N– 2,3,5,6-tetrafluorobenzoic acid**	400	Also supported by PMIRRAS analysis, –N=N-stretching at 1380 cm⁻¹ [120]
Metal oxides-aryl bonds		
U: *ITO*–**NO₂**	O1s: 532.70	[86]
E: *ITO*–**Pyrrole**	O1s: 531.5	[121]
E: *Ti*–**O–COOH**	C1s: 291.5	C1s [122]
E: *Cu*–***O*–H**	O1s: 531.8	O1s [56]
S: *MnO*–***O*–NH₂**	O1s: 533	O1s [111]

[a] E = electrochemical, S = spontaneous, C = chemical, U = ultrasonication.

Therefore, *Metal*–**aryl** and *Metal*–**N=N**–**aryl** bonds have been detected on metals, and **Metal**–***O***–**aryl** bonds on oxides; their formation will be discussed below in Section 3.7.

3.4.5 Time-of-Flight Secondary Ion Mass Spectroscopy

Time-of-flight secondary ion mass spectroscopy (TOF-SIMS) is a powerful method for investigating modified surfaces based on mass spectrometric analysis.

TABLE 3.4

Evidence for *Surface–Aryl* Bonds by TOF-SIMS

Modified Surfaces	Characteristic Peak	References Comments
$GC-C_6F_{13}$	$[CH_2-C_6H_4C_6F_{13}]^+$	CH_2 is part of the carbon surface;
	$[CH_2C_6H_4Br]^+$	$C_6H_4C_6F_{13,}$ Br are the grafted molecule [123]
$GC-[Ru(bpy)_2(apb)]$	$[C_6H_5-C_6H_4]^+$ $[(C_6H_4)_2-bpy]^+$	bpy = 2,2'-bipyridyl; apb = 4'-(4-aminophenyl) −2,2'-bipyridyl [124]
$Au-(C_6H_4)_n-H$	$[Au-C_{12}H_9]^+$	*Au* surface modified by a poly-p-phenylene film [96]
$Au(nRod)-NO_2$	$[Au-C_6H_4NO_2]^+$	[125]
$Al,$	$[Al-C_6H_4-CH_2-S-C(=S)]^+$	*AlNP* modified by a
$AlO-C_6H_4-CH_2-S-C(=S)-N-(C_2H_5)_2$	$[Al-O-C_6H_4-CH_2-S-C(=S)-N-(C_2H_5)_2]^{+-}$	photoinitiator [108]
$GC-O-C_6F_{13}$	$[O-C_6H_4C_6F_{13}]^-$	O is an oxygen from the surface [123]
$GC(OH)-N=N-C_6H_4-C_6H_5$	$[C(OH)-C-N=N-C_6H_5]^+$	This fragment corresponds to an electrophilic attack of the diazonium salt on a hydroxylated carbon ring [126]

Assignment of these spectra involves some interpretation that can be reduced by grafting, for example, **DBr**, the two isotopes of bromine, make the interpretation safer. The different TOF-SIMS analysis of diazonium-derivatized surfaces are tabulated in Table 3.4.

With this method, evidence for *GC–* and *Metal–aryl* bonds, *Metal–O–aryl* and *GC–N=N–aryl* bonds have been demonstrated, confirming the XPS results.

3.4.6 RAMAN AND INFRARED SPECTROSCOPY

In situ Raman spectroscopy was used to monitor 4-nitroazobenzene in an electrochemical cell, both as a free molecule and as a grafted monolayer on a GC electrode. The surface intensity of the Raman bands associated with the grafted phenyl $-NO_2$

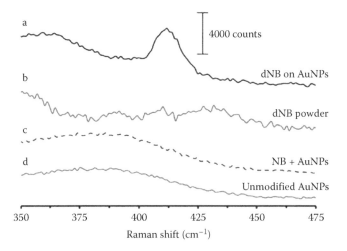

FIGURE 3.17 (a) Raman spectrum of 40 nm *Au(np)*-NO₂ showing the signature of the gold-aryl bond; (b) DNO₂ powder; (c) nitrobenzene + Au(np); and (d) unmodified Au(np). (Adapted from Laurentius, L. et al. *ACS Nano* 5, 2011: 4219–4227.)

moiety varied as the substrate was polarized between ~0 and −1.2 V/SCE, implying a strong electronic interaction between the π system of the graphitic substrate and the electrografted molecules [127]. The best proof of the existence of a bond between the substrate Au(nP) and the grafted 4-nitrophenyl group, *Au(nP)*–**NO₂**, was obtained by surface enhanced Raman spectroscopy (SERS). A small band at 412 cm⁻¹ was assigned to the *Au(nP)*–**C(aryl)** bond by careful comparison with pristine Au(nP) and with DFT calculations (Figure 3.17). The same conclusion was obtained by examination of the high resolution electron energy loss spectroscopy (HREELS) spectra [128].

Raman spectra have also been calculated for the structures of Figure 3.18 and compared with experimental spectra obtained by grafting 4-nitrobenzenediazoate on gold nanorods [125]. Structure B was found to give the best fit to the experimental

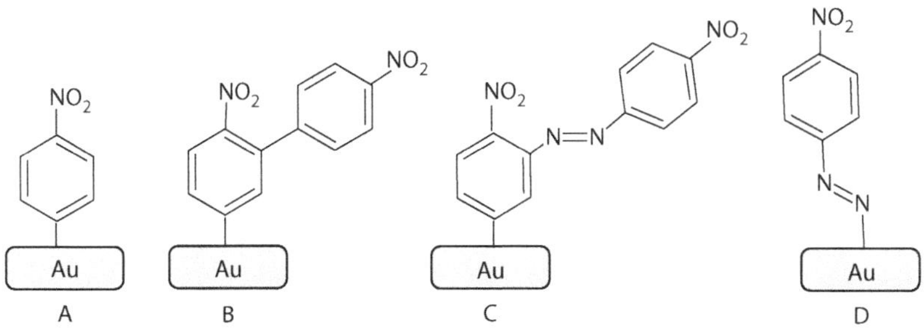

FIGURE 3.18 Different structures for which Raman spectra have been calculated.

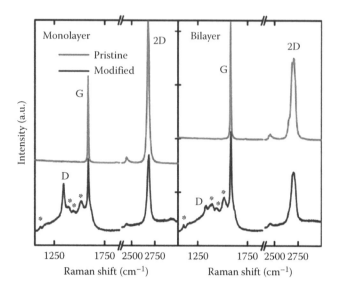

FIGURE 3.19 Raman spectra of pristine and modified mono- and bilayer graphene. Peaks introduced by nitrophenyl groups are asterisked. (Adapted from Huang, P. et al. *Acc. Chem. Res.* 46, 2013: 43–52.)

XPS Raman spectrum, and the calculated spectrum of D shows a strong band at 1763 cm^{-1} that is not found on the experimental spectrum, thus ruling out the grafting through an azo linkage from a diazoate.

Phase modulation infrared reflection absorption spectroscopy (PMIRRAS) examination of *Au* surfaces electrografted by (2,3,5,6-tetrafluoro) 4-carboxybenzenediazonium (**DF$_4$COOH**) shows azo bonds at ~1380 cm^{-1} [120]. Their presence indicates that the formation of *Au*–N=N–C$_6$F$_4$–**COOH** as multilayers does not occur due to the presence of fluorine atoms on the aromatic ring.

Raman spectroscopy is widely use for the characterization of modified graphene. The Raman spectrum of pristine graphene (and HOPG) shows two main peaks: the G band (1580 cm^{-1}), a primary in-plane vibrational mode, and the 2D band (2690 cm^{-1}), a second-order overtone of a different in-plane vibration. On modification, a new D band appears at 1350 cm^{-1}. The D mode arises from the defect involving a double resonant scattering process and is characteristic of the covalent bonds formed between graphene carbon atoms and other chemical groups. The simultaneous observation of the D band and the signature of the nitrophenyl groups is proof of the covalent bonding as shown in Figures 3.19 and 3.20 [129].

3.4.7 THEORETICAL CALCULATIONS

In parallel with experiments, theoretical calculations have been very fruitful in establishing the existence and the characteristics of a covalent bond between the surface and the aryl group. Values of bonding energies are tabulated in Table 3.5.

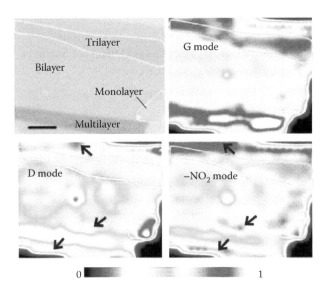

FIGURE 3.20 Raman map of **Gr-NO₂**. The top left panel shows the optical image of the investigated graphene sheet. The scale bar is 2.5 μm. Domains with different numbers of stack layers are indicated. In the 2D Raman maps of the G mode, D mode, and N–O vibration mode, the integrated intensity of each Raman mode is normalized. Note that the N–O vibration mode exhibits nearly the same distribution as that of the D mode, which proves that the nitrophenyl groups are bonded with graphene via σ-bonds. (Adapted from Huang, P. et al. *Acc. Chem. Res.* 46, 2013: 43–52.)

3.4.8 DISCUSSION

The different bonds can be summarized as follows:

1. *Direct bonding between the surface (carbon, metals, semiconductors) and the aryl group: surface–aryl.*

There is no doubt about the existence of the direct bonding on different surfaces (carbon, metals, semiconductors) which has been demonstrated both by spectroscopic methods and theoretical calculations. In particular, the observation of the **Au–aryl** bond by Raman spectroscopy nicely confirms the existence of this bond [128].

Electrochemical reduction of **DN(C₂H₅)₂** [136] and **DNO₂** [137] on HOPG has shown (by AFM) the nucleation of aryl groups at step edges and at a limited number of atomic scale defects. Further growth of the film was assigned to the formation of multilayers which covers the surface but is anchored only at these defects leading to bush-like structures. Similar nucleation was observed upon electrochemical reduction of **D–CH₂–Cl** [103]. The surface of HOPG was also patterned with **DCOOH** by scanning electrochemical cell microscopy where electrografting takes place in the meniscus at the end of a dual barrel pipette in contact with the surface. The grafted spots could be imaged (without previous rinsing) by AFM and through the Raman D band at 1350 cm⁻¹ [138]. But the films obtained on HOPG can be removed during scanning with the scanning tunnelling spectroscopy (STM) tip [137], there was

TABLE 3.5
Bonding Energies of Aryl Groups on Different Surfaces Obtained by Theoretical Methods

Modified Surfaces	Binding Energy kJ mol⁻¹	Site	References
Au–H	100	Atop	[130]
Au–H	133	Atop	[131]
Ti–H	270	Bridge	[130]
Fe–H	172	Bridge	[130]
Cu–H	117	Bridge	[130]
Pd–H	156	Atop	[130]
Si–H	293	Atop	[130]
SWCNT–H	67–168[a] 164–222[b]		[132,133]
SWCNT–Br	81–148[c] 181[d] 366[e]		[134]
Gr–H	100[f] 514[g]		[135]

[a] Depending on the tube chirality and diameter of the SWCNT, a single aryl group adsorption on metallic tubes is significantly stronger than that on semiconducting tubes, and the adsorption energy decreases with increased tube diameter.

[b] For the attachment of a second aryl group depending on their respective locations on a (8,0) SWCNT.

[c] For (20,0) and (8,0) semiconducting SWCNT.

[d] For (9,0) metallic SWCNT.

[e] For a pair of bromophenyl groups in para position on a (0,0) metallic SWCNT.

[f] For a single aryl group on a graphene sheet.

[g] For two aryl rings in para position on an aromatic ring on the graphene sheet.

therefore a doubt about a real grafting of the nitrophenyl film on the surface of HOPG in spite of the observation of a D-band [129]. The films could be anchored only at defects [136] or just deposited on the surface without any bond [137] formation.

These results can be compared with a similar report on mechanically exfoliated graphene [139]. The reaction with **DBr** provided a non-covalently functionalization of the basal region. In this case, the reaction was performed in a water/methanol mixed solvent (1:1 by volume) where the lifetime of the diazonium-derived species is very small [27–28] and the reacting species may have disappeared before the reaction is started. The result on HOPG can also be compared with that obtained on graphene where bilayers are less reactive than monolayers. For HOPG, with a large number of graphene sheets, the reactivity could be even lower [68,129,140].

The doubts about the formation of a chemical bond between HOPG and graphene and aryl groups have been alleviated by a recent paper that examined by AFM, STM, and Raman the grafting of **DNO₂** and the sterically hindered **D(3,5-*t*Bu)** (3,5-bis-*tert*-butylbenzenediazonium) [141] on HOPG and graphene [142]. These authors observed on both substrates the formation of a film both by AFM and STM and simultaneously of a D-band at ~1330 cm^{-1} (this band reflects the existence of a HOPG-aryl bond, it is small for **DNO$_2$** but increases with concentration for **D(3,5-tBu)**) and the distortion of the graphite lattice around sp^3 defects following grafting on virtually defect-free HOPG. The authors also confirmed that the film can be erased by the STM tip and that the sp^2 lattice of graphite is restored, this permits the nanopatterning of the film. Therefore, on both HOPG and graphene, there is no doubt about the grafting that is demonstrated by convergent methods electrochemistry, Raman, AFM, STM, but the film can be removed by a microscope tip [137,142] in spite of their high thermal stability [2].

Although grafting on gold has been widely investigated, there is one report where a gold surface dipped in a **DBr** acidic aqueous solution only gives a layer of ordered 4,4′-dibromobiphenyl [94]. To date, there is no rationale for these very different results on gold.

    2. *Bonding of the aryl groups on surface oxygen atoms.*

On pure oxides (MnO$_2$, Fe$_2$O$_3$) and on partially oxidized (Al, Cu, carbon) surfaces, the formation of O–aryl bonds has been established by XPS [56,86,122] and TOF-SIMS [123,126].

    3. *Bonding of aryl groups through –N=N– bonds.*

The existence of ***Surface*–N=N–aryl** bonds are more difficult to establish because azo bonds are also formed during the growth of the film. A C(OH)–C–N=N–C$_6$H$_5^+$ has been observed by TOF-SIMS [126] and assigned to an electrophilic attack of the diazonium cation on an –OH-substituted aromatic ring at the surface of carbon. This is the well-known azo coupling reaction between diazonium salts and, for example, naphthol that has been the basis of the dye chemistry. It is different from the reaction of Figure 3.1 that involves a radical.

TOF-SIMS profiles of electrografted films (starting from the diazonium salt) on copper do not show the presence of ***Cu–N=N–C$_6$H$_5$*** at the interface between the metal and the organic film. The calculated and experimental Raman spectra of ***AuNRod–NO$_2$*** (spontaneous grafting from the diazoate) do not indicate the presence of such bonds [124–125]. However, on gold after spontaneous grafting of **DNO$_2$**, a strong ***Au–N*** signal is observed by XPS (Figure 3.17c), which is attributed to the formation of ***Au–N=N–aryl*** bonds [112,131]. On vanadium [115], ***V(nP)–N=N–C$_6$H$_4$–C$_8$H$_{17}$*** structures (obtained by simultaneous reduction of VCl$_4$ and the diazonium salt by superhydride) have been proposed. However, the nanoparticles were only rinsed once in ACN so that the –N=N– bonds could also be assigned to the formation of multilayers or weakly adsorbed side products. These results indicate that ***Au–N=N–aryl*** bonds are formed on gold under electrografting and spontaneous grafting starting from the diazonium but not from the diazoate, a logical result if one assigns the formation of these bonds to the attack of the nucleophilic diazonium on the electron rich metal.

3.5   THE DIFFERENT SURFACES AND DIAZONIUM SALTS THAT CAN BE GRAFTED

The grafting of diazonium salts is a rather general method of surface modification due to the variety of diazonium salts and surfaces that can be used and to the many different grafting methods that have been described in the preceding sections.

3.5.1   THE DIFFERENT DIAZONIUM SALTS

A few diazonium salts are commercially available and 4-nitrobenzenediazonium tetrafluoroborate **DNO$_2$** has been widely used for this reason as well because $-NO_2$ is an excellent reporter group. It shows two IR signals at ~1520 and ~1340 cm^{-1} corresponding to the asymmetric and symmetric vibrations, an XPS signal at ~406 eV far from other organic nitrogens, and reversible electrochemical kinetics at E° approximately -1.20 V/SCE [2,143]. Other diazonium salts with reporting groups can be used, such as **DCF$_3$**, that show strong IR bands in the 1335 and 1230 cm^{-1} regions [54], or **DBr** that permits an easy assignment of TOF-SIMS spectra due to the two bromine isotopes [126]. Some diazonium salts recently used for surface modification are tabulated in Table 3.6. They reflect the variety of diazonium salts that can be prepared, some by simple mixing and some by more complex organic methods. They range from simple molecules to modified proteins. Substituents on the phenyl ring of the diazonium salt influence the grafting both through their electronic effects (Hammet plot, Figure 3.4) and through their steric effects.

3.5.1.1   Electronic Effects

Electronic effects are mostly observed during spontaneous grafting because the potential is determined by the reducing surface. On Zn and Ni, grafting is more efficient with the electron withdrawing $-NO_2$ substituent than with the electron donating $-N(C_2H_5)_2$ [54]. On glassy carbon, the easily reducible **DNO$_2$**, **DCF$_3$**, or **DBr** ($E_p = +0.20, -0.34, -0.35$ V/SCE, respectively) is grafted spontaneously, but not **DN(C$_6$H$_5$)$_2$**, **DNH(C$_6$H$_5$)**, **DNEt2** ($E_p^{ACN} = -0.60, -0.50,$ and $-0.56,$ respectively, on **GC**) [54,60].

3.5.1.2   Steric Effects

The steric effect of substituents has been investigated where diazonium cations are substituted with 2,6-methyl groups or a 2-ethyl group; the reaction of the radical with the surface is hindered and no grafting occurs [141,178]. With 3,5-bis-t-butyl groups, steric hindrance prevents further growth of the film and monolayers are obtained [141]. This was confirmed by (i) quantitative studies using EQCM that showed a decrease in the deposited organic layer thickness and the faradaic efficiency of the electrochemical process with the size of the 4-substituent [179], and (ii) AFM and STM on HOPG and graphene [142].

3.5.2   SURFACES THAT HAVE BEEN MODIFIED

Surface modification of nanomaterials has been the subject of much interest in recent years [180]. Some of these modifications have been performed by diazonium chemistry and include nanoparticles, nanodiamonds, fullerene, carbon nanotubes, and graphene (Table 3.7).

The modification of nanoparticles has been achieved by spontaneous grafting on a number of surfaces of metallic and metallic oxides (in this case, the bond is therefore an O–aryl bond). Modification of the surface of carbon nanotubes by diazonium salts permits the unbundling of CNTs produced by chemical vapor deposition [236],

TABLE 3.6
Diazonium Salts Used for Surface Modification

Diazonium Salts	Comments	References
DPO3H	4-Aminophenylmethylphosphonic acid was grafted on ITO surfaces by (1) electrografting of the corresponding aryl diazonium salt and (2) by self-assembly via the phosphonic acid moiety	[144]
DCOOH	Attachment of ferrocene through amide bonds	[145]
DNH2	Attachment of biomolecules, ferrocene through amide bonds	[116,146–147]
DNH2	GC–NH2 is diazotized to GC–N$_2^+$ for the attachment of:	[148]
	Nanoparticles	[149–151]
	DNA	[150]
	Enzymes	[152–153]
	Carbon nanotubes	[29,150, 154–157]
D(N2$^+$)$_2$	$^+N_2–C_6H_4–N_2^+$	[158]
DPy^{++}	Leads to GC–Py$^+$	[159]
DN3$^+$	4-Azidobenzenediazonium tetrafluoroborate for the further attachment of proteins under photo activation	[160]
D(CH2)nCH3, n = 5,13	Electron transfer through the alkyl chains	[161]
DOEG	Olygoethylene glycol diazonium salt confers proteins resistance on gold, carbon, and silicon	[162]
Glycosylated benzenediazonium		[163]
N-(4-amino-2,2,6,6-tetramethylpiperidine-1-oxyl)-4-aminobenzamide		[164]
	Oxidized to the nitroxide radical once grafted on the GC surface	

(Continued)

TABLE 3.6 (*Continued*)
Diazonium Salts Used for Surface Modification

Diazonium Salts	Comments	References
1,1,3,3-Tetramethyl-2,3-dihydroisoindol-2-yloxyl-5-diazonium tetrafluoroborate	For decorating surfaces with radicals	[165]
Calix[6]arene mono-diazonium	For the detection of uranyl	[166]
Tetradiazonium calixarene		[167]
H$_2$TPP–(N$_2^+$)$_4$		[168]
Azure A	For the catalytic oxidation of NADH	[169]

(*Continued*)

TABLE 3.6 (*Continued*)
Diazonium Salts Used for Surface Modification

Diazonium Salts	Comments	References
Azobenzene diazonium Fast Garnet GBC sulfate salt Fast black K salt		[170]
Neutral red	Electron transfer mediator to facilitate electron transfer from bacteria to electrodes	[171]
Keggin type polyoxometalate $[PW_{11}O_{39}\{Ge(p\text{-}C_6H_4\text{-}C{\equiv}C\text{-}C_6H_4\text{-}N_2)\}]^{3-}$	For the design of molecular memories	[172–173]
D-Pyrrole	On Ni	[174]
Polythiophene diazonium	For molecular junctions	[175]
Antibody–NH–C(=O)–C_6H_4–N_2^+	Immunosensing applications	[176]
Protein–NH–C(=O)–C_6H_4–N_2^+	A review	[177]

TABLE 3.7
Surfaces That Have Been Modified

Substrates	Comments	References
Carbon		
Diamond	Hydrogenated diamond can be modified by electrochemistry, spontaneously or under ultrasonication. This permits further attachment of biomolecules	[181–184]
HOPG	Grafting occurs on HOPG but the film can be removed by an AFM tip	[135–142]
GC	This is the surface that has been most modified for energy, analytical, and bioanalytical applications	[1,2,185]
Carbon fibers (CF)	For analytical purposes or carbon fiber-epoxy composites	[186–188]
Carbon black	An important material for inks, paints, fuel cells, capacitors,…	[59,189–193]
Metals		
Au	The metal of choice for fundamental investigations. See discussions on the mechanism below	[112,194]
Pt	By electrochemistry on Pt° and spontaneously on reduced Pt	[58,195–196]
Fe	Polyphenylene films confer a protection against corrosion	[88,197–198]
Cu	The electrochemical reduction of nitrate to nitrite permits the *in situ* formation of the diazonium salt and its grafting on copper	[11,55–56]
Ni		[174,199]
Stainless steel	Limitation of corrosion and temperature resistance of the coating	[107,114]
Nitinol	An attempt to limit the release of Ni_2^+ but DCOOH is not efficient	[200]
Semiconductors		
Si	On hydrogenated Si (SiH) the polyphenylene film passivates the surface. The reaction can be spontaneous or electrochemical	[100,201–202]
SiGe	Spontaneous reaction	[73]
Ge	Permits the spontaneous passivation of the surface by the polyphenylene film	[74,75]
GaAs		[203]
Oxides		
ITO	Attachment of proteins, DNA	[204–206]
SiO_2	For further photopolymerization	[89]
TiO_2, Fdoped-SnO_2	Both spontaneously and by electrochemistry	[122,207–208]
Polymers		
PMMA		[90]

allowing separation of metallic from semiconducting CNTs [237–238]. However, this surface modification by aryl groups decreases the conductivity of the CNT and the restoration of unmodified CNT can only be achieved at high temperature [105–106]. The surface modification of graphene by diazonium salts has been reviewed (Table 3.8). One of the main objectives of these modifications is opening a bandgap in its gapless band structure, as well as improving the processability of this intrinsically insoluble material. The reactivity of graphene with diazohydroxides depends on the number of layers. For example, single graphene sheets are found to be almost 10 times more reactive than bi- or multilayers of graphene according to the relative disorder (D) peak in the Raman spectrum examined before and after chemical reaction in water. The

TABLE 3.8
Surface Modification of Nanomaterials

Substrates	Comments	References
Nanoparticles		[6]
Au(nP)	DNA sensors	[209–210]
Au(nRod)	Aryl groups can partly replace the stabilizing CTAB surfactant	[125]
Ag(nP)	Antibacterial properties of water-dispersible silver nanoparticles	[115]
γ-*Fe$_2$O$_3$–SiO$_2$(nP)*	For hyperthermia	[211]
Co@C(nP)	For imaging	[212]
Fe$_3$O$_4$(nP)	Magnetic *nP*	[213]
Al$_2$O$_3$(nP)	Energetic materials	[108]
Nanodiamonds		[214–215]
Highly boron doped diamond BDD	Protein–polymer bioconjugate	[216]
Detonation diamond	Post modification through Suzuki coupling	[217]
Fullerenes		
C-60	Attachment to Si surface	[218]
CNT		[63,219–223]
	Electrochemical grafting of buckypaper	[224–225]
	Electrochemical modification of a single SWCNT	[226]
	Fe as reducing agent	[227]
	Spontaneous modification in ACN	[228]
	Spontaneous modification in aqueous solution	[105]
	Spontaneous modification in pyrrolidone	[229]
Graphene		[66,70,129,230–234]
	Localized grafting of graphene oxide	[235]
	Distinction between single and bilayer graphene	[68,129,140]
	Patterning a film of D(3,5-*t*Bu) with an AFM tip	[142]

reactivity of edges is at least two times higher than the reactivity of the bulk single graphene sheet [68,129,140,239]. The difference in reactivity between mono and bilayers of graphene can be observed in Figure 3.19, where the D band is much smaller in the bilayer. Figure 3.20 shows the Raman maps [129] of a modified graphene sheet. The intensity map of the G mode indicates the distribution of the numbers of the stack layers on graphene flakes; that is, more layers exhibit stronger G mode intensity. The distribution in the intensity maps for the D mode and the nitrophenyl peaks are nearly the same but inhomogeneous indicating the inhomogeneous formation of covalent bonds. In addition, more grafting appears on the edges, and on monolayers more than bilayers. An interesting difference [142] is shown by the comparative grafting of **DNO₂** and **D(3,5-*t*Bu)** on the surface of HOPG and graphene (Figure 3.21). Grafting of **DNO₂** (as reported in Section 3.4.8) on defectless HOPG or graphene leads to a limited number of "bushes" of poly 4-nitrophenyl groups (that can be observed by AFM or STM, and correspond to a small D band that does not increase with the thickness of the "bushes"). On the contrary, the sterically hindered **D(3,5-*t*Bu)** [141] (see Section 3.6.2) provides a monolayer (and the D band increases with the concentration).

In summary, a large number of diazonium salts starting from very simple ones to diazotized proteins have been bonded to a large number of surfaces and the present state of the art opens the way to numerous other examples to be discovered.

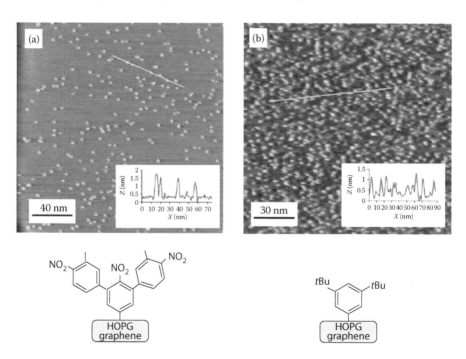

FIGURE 3.21   STM images of HOPG modified by (a) **DNO₂** and (b) **D(3,5-*t*Bu)** ($c = 1$ mM). The insets show line profile; with **DNO₂** isolated "bushes" ($d = 2.7$ nm) of polyphenylene film are observed with th ~ 1.5 nm, for **D(3,5-*t*Bu)** a monolayer of closely spaced spots with th ~ 0.7 nm; at $c = 10$ mM complete coverage of the surface by the monolayer is observed. (Adapted from Greenwood, J., et al. *ACS Nano* 9, 2015: 5520–5535.)

3.6   CHARACTERIZATION OF THE FILM

In the preceding section, we described the bonding of the film on surfaces. We now examine the characterization of the film in terms of thickness, structure, surface concentrations, and the influence of the different grafting conditions on these parameters. One of the most important parameters for describing these films is its thickness. We will examine how the thickness of polyphenylene films can be controlled from submonolayers to micrometer thick films. Indeed, at the different thiol SAMs that always give monolayers, attachment of aryl groups at surfaces by diazonium chemistry involves the formation of radicals that can attack the first grafted groups to give polyphenylene films. Therefore, efforts have been made either to obtain submonolayers, monolayers of nanometer thickness or quite thick layers of micrometer thickness. In the following, the films will be characterized through their dry-state thickness (th, nm), their surface concentration Γ_{surf} or Γ_{vol} (mol cm^{-2}), whether it concerns the surface concentration of groups directly attached to the surface or the total number of repeating groups in the entire thickness of the film. As a simple means of comparison, for a close-packed film of nitrophenyl groups that cannot rotate and irrespective of their bonding to the substrate, molecular models give $\Gamma_{surf} = 12 \times 10^{-10}$ mol cm^{-2}. This gives an upper limit for the surface concentration of a monolayer.

3.6.1   SUBMONOLAYERS

Anthraquinone-2-diazonium tetrafluoroborate has been adsorbed onto an edge plane pyrolytic graphite electrode to form a thin unreacted submonolayer film. After transfer to a buffer solution containing no diazonium salt, the adsorbed material was thermally decomposed at room temperature and a grafted anthraquinone film was obtained with a surface coverage $\Gamma_{surf} < 2 \times 10^{-10}$ mol cm^{-2} [240]. Nitroazobenzene submonolayers have been obtained by spontaneous modification of PPF surfaces. The apparent thickness was th = 0.6 nm and Raman spectra confirmed the structure of the film [241]. On gold, 0.06 $\Gamma_{monolayer}$ was found, as determined by integration of the voltammogram of the nitrophenyl group [44]. Submonolayers of the bulky inorganic polyoxometalate were obtained both by preadsorption, and spontaneous and electrochemical grafting with surface concentrations down to $\Gamma_{surf} = 0.1 \, \Gamma_{monolayer}$ [173].

3.6.2   MONOLAYERS

It is possible to control the thickness of the polyphenylene films by controlling (i) the time duration of the reaction, (ii) the concentration of the diazonium salts, or (iii) the potential(s) of the electrolysis or cyclic voltammetry. The first two methods apply to spontaneous and chemical grafting, while all the three apply to electrografting. Using these parameters, monolayers have been attached to PPF for microelectronic applications. For example, a film of *PPF*–N=N–C$_6$H$_4$–NO$_2$ groups has been obtained by electrochemical reduction (one cyclic voltammetry scan between +0.4 V and 0 V/ Ag/Ag$^+$, scan rate $v = 0.2$ V s^{-1}) of the corresponding diazonium salt. A thickness of

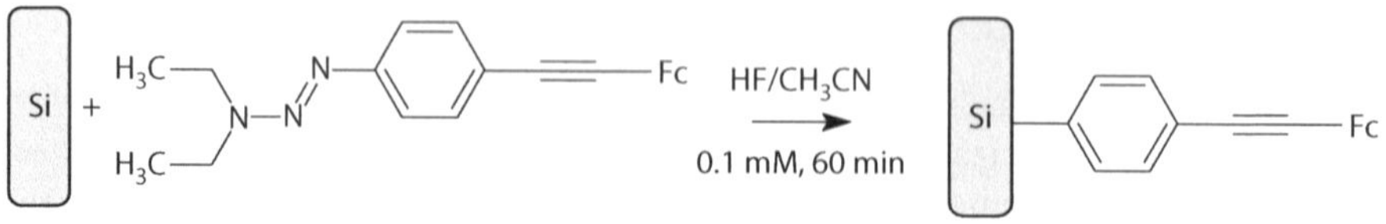

FIGURE 3.22 Grafting of a triazene on Si to form a monolayer of ferrocenylethynylphenyl groups.

1.0 nm was measured by AFM scratching, in good agreement with results from a molecular model ~1.2 nm [242]. On Si(100), a monolayer of phenylethynylferrocene (0.99 nm long) (Figure 3.22) with a thickness of th = 0.8 nm was obtained by spontaneous grafting upon varying the concentrations and reaction times [243].

However, these trial and error methods are time consuming and more direct methods have been devised. The first method uses steric hindrance to prevent further radical attacks on the first grafted aryl groups. For example, by electrografting **D(3,5-*t*Bu)**, a 1.1-nm film was obtained that is very close to a monolayer [141]. However, the two bulky groups prevent further modification of the film. This becomes possible using a 4-trimethylsilyl protecting group as shown in Figure 3.23. After deprotection in the presence of fluoride ions, an ethynylphenyl group remains on the surface [244,245] that can be further modified by "click" chemistry [246].

The second method uses a "formation-degradation approach" that is summarized with three examples in Figure 3.24. In these examples, the bulky substituents partially protect the first grafted aryl group from further radical attack and are then cleaved to give a monolayer. In Figure 3.24(a) the diazonium salt of a disulfide provides a multilayer th = 3 nm. After electrochemical reduction of the disulfide, a near-monolayer th = 1.5 nm film is obtained [247]. In (b), a diazonium group substituted by an hydrazone group is electrografted (th = 1.9 nm) and the cleavage in acidic medium provides a near-monolayer benzaldehyde film (th = 1.2 nm) [248]. In (c), a diazonium salt with an ester group is electrografted (th = 3.1 nm) and then hydrolyzed to a monolayer of phenylcarboxylic groups (th = 0.4 nm) [249].

The third method makes use of radical traps; it is the most general method because it can be applied to any diazonium salt. For example, on a carbon electrode,

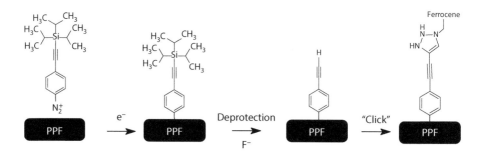

FIGURE 3.23 Trimethylsilyl protecting group for the formation of monolayers.

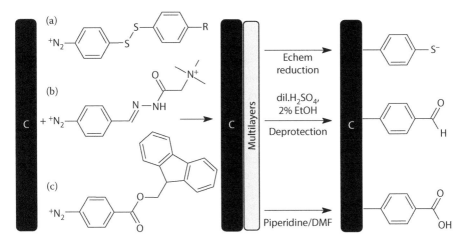

FIGURE 3.24   Formation–degradation approach for the formation of monolayers.

using one CV cycle between ~+ 0.5 and −0.6 V/SCE in a 1 mM solution of **DNO2**, a film of polynitrophenylene was obtained corresponding to $\Gamma_{vol} = 17 \times 10^{-10}$ mol cm^{-2} (measured by integration of the cyclic voltammogram of the nitrophenyl group). Upon addition of 2 mM of the radical trap DPPH (2,2-diphenyl-1-picrylhydrazyl), it decreases to $\Gamma_{surf} = 6.4 \times 10^{-10}$ mol cm^{-2}, a value corresponding to a monolayer. This is also exemplified by EQCM in Figure 3.25, where the mass deposited on the electrode was found to decrease to that of a monolayer [250].

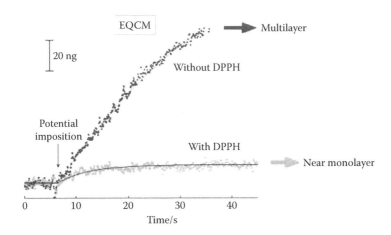

FIGURE 3.25   Mass variation versus time on carbon covered quartz measured by EQCM for the electrochemical grafting using **DNO$_2$** (1 mM) at a fixed potential of −0.5 V without DPPH and with 1 mM DPPH. Data are fitted using the Langmuir model (solid lines). (Adapted from Menanteau, T., E. Levillain, and T. Breton. *Chem. Mater.* 25, 2013: 2905–2909.)

As discussed previously, the structure of these monolayers has been investigated. The estimated maximum surface concentration of a compact nitrophenyl monolayer layer is 12×10^{-10} mol cm^{-2} compared to $\Gamma_{surf} = 6.4 \times 10^{-10}$ mol cm^{-2} from EQCM and $\Gamma_{surf} = 2.5 \times 10^{-10}$ mol cm^{-2} from combined AFM and electrochemical experiments in ACN [251] for a film with thickness equal to that of a monolayer. On H–Si(111), a monolayer has been characterized both by STM ($\Gamma_{surf} = 5 \times 10^{-10}$ mol cm^{-2}) and Rutherford back scattering (RBS) ($\Gamma_{surf} = 6.7 \times 10^{-10}$ mol cm^{-2}) [252]. These values are quite similar considering the different materials (i.e., carbon- covered quartz in the first case, a flat PPF electrode in the second, and monocrystalline Si in the third). These results can be considered as a basis for further discussion on surface concentrations of aryl groups. The stereochemistry of aryl groups has also been determined. The calculated angle (by DFT) between the surface normal and the plane of the phenyl ring for *Au*-C_6H_5 was found to be $\Theta = 38°$ [130]. A similar value was determined by IR spectroscopy [242] as $\Theta = 31°$ for *PPF*-C_6H_4–N=N–C_6H_4–NO_2. The nitrophenyl groups were found in upright position in *Diamond*-NO_2 by XPS [253]. Finally, there is only one example where an organized array of grafted aryl groups (*Si*–Br) has been observed on H–Si(111) by STM [252].

3.6.3 THIN LAYERS

Electrografting of diazonium salts generally provides thin layers with thicknesses ranging from monolayer to less than 50 nm, while thinner films ($\leq$10 nm) are obtained by spontaneous grafting. For example, film (*Si*–COOH) thicknesses ranging from 1.9 to 4.6 nm were obtained on different types of Si [73]. A schematic representation of these bonded oligomers is given in Figure 3.26.

3.6.3.1 Structure of the Films

The surface bond was discussed in Section 3.4. The oligomer that is formed on the surface has a polyphenylene structure with some of the aryl groups bonded by azo groups. No order could be found as in SAMs having different thiol groups. However, by first attaching a thin layer of phenyl groups on the surface of gold and then growing a polyphenylene layer by electrochemical oxidation of biphenyl, it was possible to obtain a regular poly(para-phenylene) film (as shown by the characteristic IR spectrum) strongly bonded to the surface as shown in Figure 3.27 [96].

We now examine the structure of these oligomers and some of their properties. The presence of aryl groups has been identified by the characteristic IR and Raman bands of the aryl group itself and its substituents. For example, the vibration of the aryl ring is observed at 1605 cm^{-1} in the spectra of *Au*–COOH and 1630 cm^{-1}. In contrast, for *Au*–C_6F_4–COOH [120], the C–F bands appear in the 1273–1203 cm^{-1} range and the –COOH substituent is observed at 1720 cm^{-1}. The nitrophenyl group has often been used to ascertain the presence of a film on a surface through the two asymmetric and symmetric stretching bands of the –NO_2 group at ~1520 and ~1340 cm^{-1} [63], through the N1s peak at 406 eV, through the reversible electrochemical signal at approximately −1.2 V/SCE [254], and also through its TOF-SIMS signature [126]. The presence of azo bonds in a film has been characterized by XPS [56,112], IR at 1380 cm^{-1} [120], and TOF-SIMS [126].

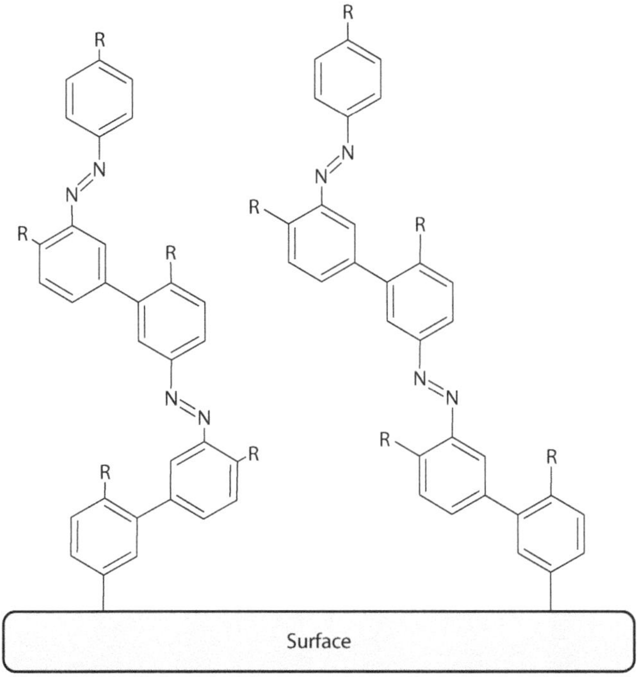

FIGURE 3.26   Schematic representation of a polyphenylene structure bonded to a surface by reduction of a diazonium salt.

3.6.3.2   Surface Concentration

The surface of concentration of the films can be determined by a limited number of methods. RBS is probably the most reliable method because it gives the number of atoms cm^{-2} of a reporter atom directly [252] (with a mass larger than that of the substrate; for example, Br, I). However, the most frequently used method involves integration of the voltammetric peak of a reporter group (e.g., the reduction of the nitrophenyl group to its radical anion in ACN). In this method, $Q_{NO2} = n F \Gamma_{vol}^{NO2} S$, where Q_{NO2} is the charge in coulombs (C) calculated from integration of the voltammetric peak of the nitrophenyl group, where $n = 1$ is the number of electrons, F the Faraday, Γ is the surface concentration in mol cm^{-2}, and S the surface area in cm^2. For 4-nitrophenyl groups, the voltammogram corresponds to a 1e$^-$ reversible electron transfer ($E°$ approx. -1.2 V/SCE) in aprotic medium, while in aqueous medium a 4e$^-$ reduction leads to phenylhydroxylamine (E approx. -0.7 V/SCE); on the reverse scan of the reversible system, hydroxylamine/nitroso is observed ($E°$ approx. -0.3 V/SCE) (Figure 3.28a and b [195,255]). This method is limited to very thin films (<3–4 nm). For thicker films, some nitrophenyl groups become electrochemically "silent," probably due to the lack of solvation and the absence of counter anions in the film. This has been shown through graphs of surface concentration (measured as above) versus thickness (by AFM scratching). The plot is linear up to 2 nm for thin films. For thicker films, the surface concentration measured by

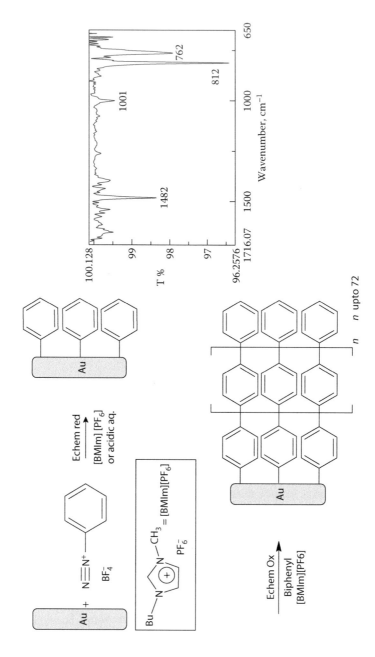

FIGURE 3.27 Procedure for producing regular PPP films covalently attached to a gold substrate. (Adapted from Descroix, S. et al. *Electrochim. Acta* 106, 2013: 172–180.)

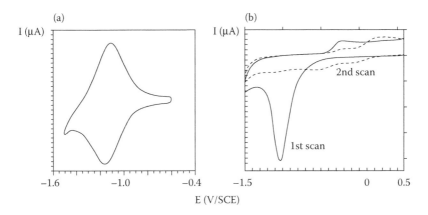

FIGURE 3.28 Cyclic voltammograms of the nitrophenyl group in (a) *Ni*-**NO**$_2$ in ACN; (b) *GC*-NO$_2$ in 0.25 M H$_2$SO$_4$. (Adapted from (a): Bernard, M.-C. et al. *Chem. Mater.* 15, 2003: 3450–3462, (b): Yu, S. S. C. et al. *Langmuir* 23, 2007: 11074–11082.)

electrochemistry becomes constant with increasing thickness [251,256–257] (Figure 3.29).

3.6.3.3 Electrochemical Activity of Nitrophenyl Groups

The question addressing electrochemical activity of the different nitrophenyl groups was explored [258] by growing the different structures shown in Figure 3.30. In (A), the nitrophenyl film is separated from the electrode by a non-electrochemical active layer. Thus, the electron transfer is considerably slower for the outer layer, but the groups in this outer layer are solvent and electrolyte accessible. In (B), the presence of a thick non-polar polyphenylene (**GC–H**) top layer slows down electrolyte permeation in the inner part of the nitrophenyl film without affecting the overall coverage. (C) represents an 11-nm thick nitrophenyl layer where some of the groups are electrochemically inactive. These inactive groups are believed to be in the bulk of the film. Multiple cyclic voltammograms were shown to permit removal of deposited material, thus increasing the porosity and creating pinholes in the film.

3.6.3.4 Thickness of the Films and Deposited Mass

Thickness can be measured by ellipsometry, by profilometry, by infrared spectroscopy if a calibration curve has been established, and by AFM scratching. The thickness is correlated to the surface concentration, while the mass can be obtained with EQCM. Figure 3.29b and c shows the thickness of **PPF–NO$_2$** measured by AFM scratching. Figure 3.31 shows that the mass deposited on the electrode increases as the potential becomes more negative [259]. This result is in agreement with thickness measurements obtained:

a. With **PPF–C$_6$H$_4$–N=N–C$_6$H$_4$–NO$_2$**: After a single voltammetric scan to +0.22, +0.02, −0.12, −0.22, and −0.42 V versus SCE, different films were obtained with thicknesses of 1.0, 1.9, 2.1, 2.3, and 2.6 nm, respectively [260].

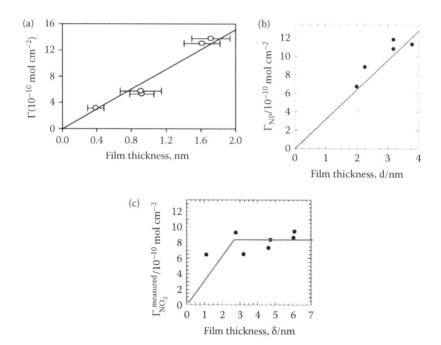

FIGURE 3.29 Plots of surface concentration measured by electrochemistry versus film thickness measured by AFM scratching (a) for *Au*-NO$_2$ films spontaneously formed on gold; (b) *PPF*-NO$_2$. • in ACN, ○ in aqueous solution; (c) *PPF*-C$_6$H$_4$–N = N–C$_6$H$_4$–NO$_2$. (Adapted from (a): Lehr, J. et al. *Langmuir* 25, 2009: 13503–13509, (b): Brooksby, P. A. and A. J. Downard. *Langmuir* 20, 2004: 5038–5045, (c): Brooksby, P. A. and A. J. Downard. *J. Phys. Chem. B* 109, 2005: 8791–8798.)

b. On *GC*: 10 min chronoamperometry was performed at E approximately −0.6, −0.8, and −1.2 V/SCE. The surface concentration increased continuously, $\Gamma = 2.1$, 2.4, and 4.2×10^{-10} mol cm^{-2}, respectively [35].

c. On *GC*: The charge transfer resistance of the film (qualitatively related to its thickness) increases more than tenfold by shifting the potential from the voltammetric peak potential to a value 300 mV more negative [9].

All of these results are in clear contradiction with the result of Figure 3.13. A possible explanation for the continuous growth of the film as the potential is pushed

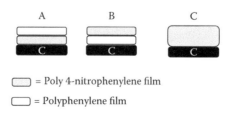

FIGURE 3.30 Illustration of electrochemically active nitrophenyl groups.

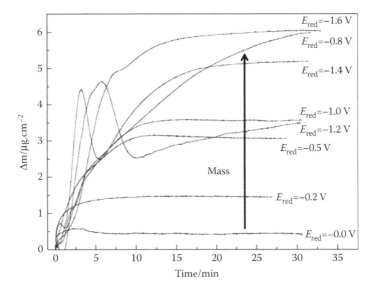

FIGURE 3.31 Mass of **DNO$_2$** deposited by EQCM in ACN on an Au electrode. (Adapted from Haccoun, J. et al. *Progress Org. Coatings* 63, 2008: 18–24.)

to more and more negative values past $E°(R\cdot/R^-)$ could be that the potential at the external surface of the already grafted film must be less negative than the potential imposed to the electrode due to the ohmic drop in the film. Needless to say, more experiments are necessary to clarify the variation of the mass and thickness as a function of potential. This is a very important point in the electrografting of diazonium salts.

Ellipsometry permits the thickness of the film to be monitored in real time and *in situ* during the chronoamperometric reduction of **DNO$_2$** (Figure 3.32a). A good correlation of thickness with charge used for reduction and the presence of nitrophenyl groups on the surface at the end of the experiment (IRRAS spectrum) is observed. Figure 3.32b represents the faradaic efficiency uniformly distributed on the Au-electrografted surface [261].

The film also depends on the medium in which they are prepared. For example, with **Au–NO$_2$**, grafting is much less efficient in aqueous solution than in ACN as shown by EQCM [179] (Figure 3.33). In addition, these films swell when passing from aqueous to organic solution and shrink under the reverse change [251,256].

The films can also have pinholes, which have been observed by AFM. Figure 3.34 shows an AFM image of a **Diamond–NO$_2$** film obtained by a single voltammetric scan. Protrusions up to 3 nm are observed which indicates that some locations on the surface were not modified, while multilayers were formed at other places [253] (see also Figure 3.13).

3.6.3.5 Redox Probes

The compactness of the film (in particular with or without the presence of pinholes) has also been qualitatively estimated through the use of redox probes, where

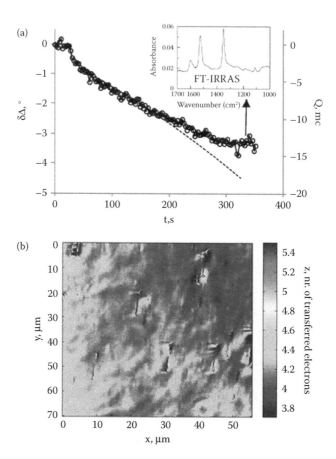

FIGURE 3.32 Real-time and *in situ* ellipsometric monitoring of NBD electrografting on an Au surface. (a) Variation of thickness, th, correlated with charge (dashed line) for a chrono-amperometric reduction of **DNO₂**. The inset shows a FT-IRRA spectrum recorded *ex situ* on the grafted surface. (b) Distribution of the faradaic efficiency (the small difference observed in the right is an instrumental artifact). (Adapted from Munteanu, S. et al. *Anal. Chem.* 85, 2013: 1965–1971.)

the voltammogram of a reversible system (ferrocene, $Fe(CN)_6^{3-}$, $Ru(NH_3)_6^{3+}$, etc.) is recorded on a carefully rinsed modified electrode. If the grafted film hinders or prevents the access of the redox probe to the electrode, the voltammogram will be modified or suppressed (Figure 3.35a and b, respectively [170,262]). However, such voltammograms are difficult to interpret, even qualitatively, because the blocking efficiency of the film depends on a number of parameters such as:

a. *Thickness*. The voltammogram of tri-*p*-tolylamine is nearly unaffected (by comparison with that recorded on a bare GC electrode) on a "thin" **GC–CH₂–COOH** film, but disappears in the presence of a "thick" film (Figure 3.35c).

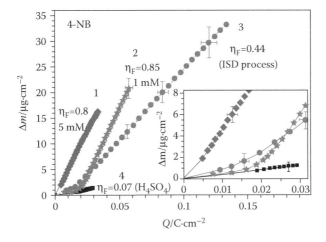

FIGURE 3.33 Mass change, Δm, as a function of the charge obtained from EQCM measurements. *Au*-NO_2 obtained from **DNO_2** in ACN (**1**) 5 mM; (**2**) 1 mM; (**3**) **DNO_2** prepared *in situ* (1 mM) in 0.1 M H_2SO_4; (**4**) 5 mM. η_F are the faradaic efficiencies. (Adapted from Zhang, X. et al. *Z. Phys. Chem.* 228, 2014: 557–573.)

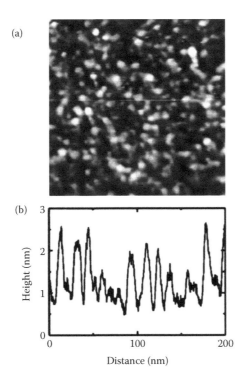

FIGURE 3.34 (a) AFM image and (b) profile of a ***Diamond***-NO_2 film obtained by a single voltammetric scan. (Adapted from Uetsuka, H. et al. *Langmuir* 23, 2007: 3466–3472.)

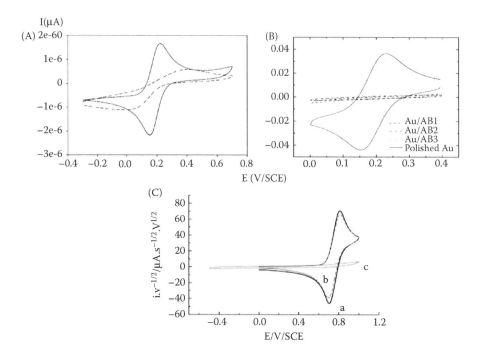

FIGURE 3.35 Redox probe experiments. Cyclic voltammograms of unmodified (A) GC and (B) Au electrodes in aqueous $Fe(CN)_6$ solution (solid line) and the same electrodes modi-fied by (A) cyclodextrin and (B) azobenzene. (C) Voltammogram of tri-p-tolylamine on (a) a bare GC electrode, (b) a "thin," (c) a "thick" **GC-CH₂COOH**. In (A) the sigmoidal shape is assigned to the presence of pinholes in the film. (Adapted from (A) Diget, J. S. et al. *Langmuir* 28, 2012: 16828–16833, (B) Kibena, E. et al. *Chem. Phys. Chem.* 14, 2013: 1043–1054, and (C): Noël, J.-M. et al. *Langmuir* 25, 2009: 12742–12749.)

b. *Chemical nature of the grafted layer.* The voltammogram of dopamine is nearly unaffected by the presence of a **GC–naphthyl**, but disappears on **GC–NO₂** [263].

c. *Charge of the layer.* **GC–SO₃⁻** in aqueous pH = 7.4 solution completely blocks the voltammogram of $Fe(CN)_6^{3-}$ but not that of $Ru(NH_3)_6^{3+}$.

d. Hydrophilicity/phobicity of the film [264].

e. *Substrate itself.* The voltammogram of $Fe(CN)_6^{3-}$ is blocked on **GC–SO₃⁻** but not on **Au–SO₃H** [265].

f. *Redox probe.* The voltammogram of tri-p-tolylamine can be observed through a **GC–NH₂** film with a sigmoidal shape that indicates a direct reduction inside pinholes. However, on the same film the bulky trans[-(η²-dppe)₂Ru(–C≡C–Ph)₂] does not show any voltammogram [266].

Using SECM, it was possible to obtain a more quantitative estimation of the pin-holes on an **Au–CH(CH₃)–Br** film. Aqueous redox probes such as $Fe(CN)_6^{3-}$ would penetrate 50 nm macropores while organic ones such as terephtalonitrile could

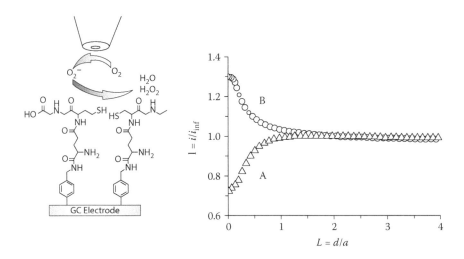

FIGURE 3.36 SECM approach curves (Au electrode, $a = 5$ μm) on a GC modified electrode by glutathione (GSH). FcMeOH (10^{-3} M) is used as a redox probe. A: before O_2 reduction; B: after 10 s of O_2 reduction. (Adapted from Latus, A. et al. *Langmuir* 27, 2011: 11206–11211.)

be reduced inside 5–10 times smaller mesopores or nanopores [267]. SECM has also been used to open pinholes in the grafted layer [268]. A GC surface modified with glutathione by diazonium chemistry (Figure 3.36, triangles) shows negative feedback, indicative of slow charge transfer between the mediator FcMeOH$^+$ and the modified surface. This behavior indicates that the layer has a strong blocking character, although not sufficient for totally inhibiting permeation of the mediator. After 10 s of reaction with superoxide, a new approach curve (Figure 3.36, circles) is recorded. Positive feedback is observed, indicating that FcMeOH$^+$ now permeates the layer. The presence of new charge-transfer pathways in the layer is likely due to the transformation of GSH to the corresponding disulfide GSSG.

Very interesting behavior toward redox probes is demonstrated by the thin films obtained from 1-(2-bisthienyl)-4-aminobenzene and 1-(2-thienyl)-4-aminobenzenediazonium (Figure 3.37, Cb and Cc, respectively). When decamethylferrocene ($E° = -0.21$ V/SCE) is used as a redox probe, its reduction is shifted to more negative potentials and becomes irreversible (Figure 3.37a); that is, the layer behaves as a diode. When the redox probe is thianthrene ($E° = 1.13$ V/SCE), both its reduction potential and its reversibility are maintained, and the films are now transparent to electron transfer (Figure 3.37b) [269]. The same phenomenon is observed with thin films obtained from the 2-terthiophenyldiazonium cation [270].

3.6.3.6 Kinetics of Grafting

Grafting kinetics have been investigated for submonolayers of **Au–NO₂** by EQCM [44], assuming preadsorption of the diazonium salt before the grafting step. The kinetics were shown to be concentration dependent; for example, $k_{obs} = 5.5–10.2 \times 10^{-3}$ s^{-1} when concentration increases from 5 to 100 μM.

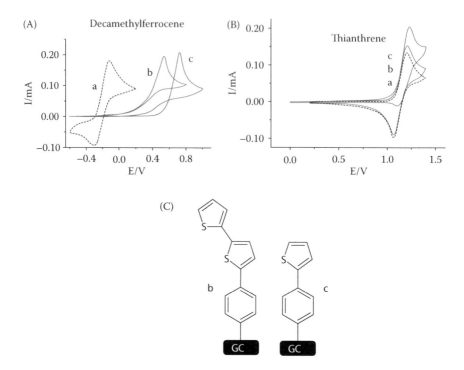

FIGURE 3.37 Cyclic voltammograms of decamethylferrocene (A) and thianthrene (B) on thiophene modified electrodes (Cb and Cc). (Adapted from Fave, C. et al. *J. Phys. Chem.* 112, 2008: 18638–18643.)

3.6.3.7 Electron Transfer Rates

Electron transfer rates through the films have been measured by attaching a redox species to the film, as summarized in Table 3.9. The rates are seen to vary over a wide range. When ferrocene is directly attached to a surface through a $-C_6H_4-C(=O)-NH-CH_2-$ or $-C_6H_4-C\equiv C-$ tether, the rates are relatively high

TABLE 3.9
Electron Transfer Rate Constants

Substrate-Aryl Groups	Rate Constants	References
GC–CONH–horseradish peroxidase	$k_s = 10.3\ s^{-1}$	[271]
GC–CO–NH–C_6H_4–C≡C–C_6H_4–POM$^{4-/3-}$	3, 0.3, and 25 s^{-1}depending on $\Gamma = 1.3, 2, 2.5 \times 10^{-11}$ mol cm^{-2}	[173]
GC–C_6H_4–C≡C–C_6H_4–POM$^{4-/3-}$ POM=polyoxometalate	2×10^{-4} cm s^{-1} to be compared with $k_{ET,sol} = 0.064$ cm s^{-1} for the same species in solution	[172]
GC–Os(bpy)$_3^{2+/3+}$	11.4 s^{-1} in ACN and 5.4 in PBS buffer	[272]
GC–CO–NH–CH_2Fc	17 s^{-1}	[145]
Si–C≡C–Fc	164 s^{-1}	[243]

but decrease to smaller values with polyoxometalates (POMs). The rate constant measured for the same POM decreases by more than 300 in going from a diffusing to an attached species. With submonolayers of POM, charge transfer from the electrode to the POM is the kinetically limiting (hopping) process involving an electrostatic contribution, likely due to charge transfer to the negatively charged POM structure.

Phenyl monolayers on gold have been used with much success for organic electronic applications (a thin polyphenylene film is included between two metallic or carbon contacts as shown in Figure 3.38a). The mechanisms of electron transfer through these layers is therefore of prime importance [273]. Three distinct transport mechanisms have been observed for 4.5–22 nm thick oligo(phenylbithiophene) layers between carbon contacts (Figure 3.38a): (i) tunneling is operative when $d < 8$ nm, (ii) activated hopping when $d > 16$ nm for high temperatures and low bias, and (iii) when $d = 8$–22 nm, a third mechanism operates consistent with field-induced ionization of highest occupied molecular orbitals or interface states to generate charge carriers.

The J/V curves [274–275] of these oligopolyphenylene films are shown in Figure 3.38b for different oligophenylene films. The curves show no hysteresis, are independent of scan rate (from 1 to 100 V s^{-1}) and may be repeated thousands of scans without observable change, indicating the very high stability of these devices.

3.6.3.8 Carbon Nanotubes

Chemical formation of variable thickness layers has also been achieved by increasing the molar ratio between **DOCH$_3$** and pristine MWCNT (Figure 3.39). In this way, it was possible to change the solubility of the nanotubes from mainly soluble in DMF (a typical solvent of CNT) to mainly soluble in EtOH (a solvent of the polymethoxyphenylene layer) [229].

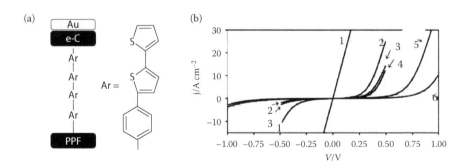

FIGURE 3.38   (a) An example of a molecular solid state junction used for the determination of charge transport mechanisms. (b) J/V curves for molecular solid state junction PPF/Ar/ Cu: **1**: PPF/Cu; Ar = **2**: 4-nitrobiphenyl (th = 2.0 nm), **3**: biphenyl (th = 1.5 nm), **4**: fluorenyl (th = 1.4 nm), **5**: biphenyl (th = 1.9 nm), **6**: 4-nitrobiphenyl (th = 3.7 nm). (Adapted from a: Yan, H. et al. *Natl. Acad. Sci. USA* 110, 2013: 5326–5330. b: McCreery, R. L. et al. *Chem. Phys. Chem.* 10, 2009: 2387–2391.)

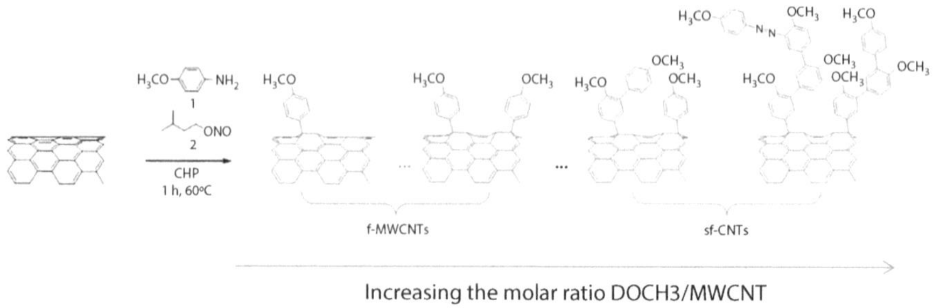

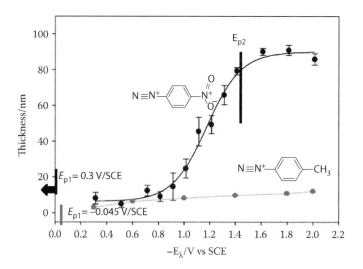

Increasing the molar ratio DOCH3/MWCNT

FIGURE 3.39 Modification of the surface of MWCNT from low to high surface concentration. (Adapted from Salice, P. et al. *Carbon* 74, 2014: 73–82.)

3.6.4 THICKER LAYERS

Thicker layers can also be obtained from diazonium salt. Under usual conditions, the growth of the film stops because the electrons cannot reach the outer surface of the film to reduce new diazonium cations. This situation can be reversed if an electron carrier such as a reversible couple is imbedded in the film and activated. This was achieved during the formation of **AuNO2** when the potential of the electrode was maintained at a potential sufficient to reduce the nitrophenyl groups (Figure 3.40); a thickness of 86 nm was obtained [42]. High thickness films were also obtained with anthraquinone groups [276–277] and with a Ru(bipy)$_3$ complex [95]. An important

FIGURE 3.40 Film thickness of **Au-NO$_2$** and **Au-CH$_3$**, electrografted through 10 consecutive cyclic voltammetric cycles with varying values of the negative limit (E_λ) of the scan. The peak potentials corresponding to the reduction of the diazonium salt (E_{p1}) and the nitro groups (E_{p2}) are indicated approximately. (Adapted from Ceccato, M. et al. *Chem. Mater.* 23, 2011: 1551–1557.)

point is that with **DNO₂** and **DAQ**, thick films could only be obtained by repetitive voltammetric scans. For example, with **DAQ**, 138 and 1.5 nm films could be obtained after 50 scans and chronoamperometry, respectively. This was assigned to the removal of any deposited material (such as dimers, etc.) and creation of permeation channels in the film by sweeping the potential. The formation mechanism of these thick films was investigated by EQCM. A mass increase was observed even during the negative scan and was assigned to the attachment of new anthraquinone groups and to the reversible association of counter cations from the supporting electrolyte.

3.6.5 MIXED LAYERS

Mixed aryl layers have also been obtained from diazonium salts, that is, by reduction of a mixture of two diazonium salts [145,278]. A binary mixture of aryl diazonium salts bearing oppositely charged para-substituents (**D–SO₃⁻** and **D–N(CH₃)₂⁺**) has been electrografted [279].

Interestingly, in the mixture solution, the two aryl diazonium salts undergo reductive adsorption at the same potential, which is distinctively less negative than the potential required for the reduction of either of the two aryldiazonium salts alone. Additionally, the surface ratio of the two phenyl derivatives is consistently 1:1 regardless of the ratio of the two aryl diazonium salts in the modification solutions. The grafting of the bithiophene phenyl (BTB) diazonium salt complexed by β-cyclodextrin (CD) provides a layer of grafted BTB oligomers complexed by CD. When CD is removed from the surface, pinholes are created and **DNO2** can be electrografted to give a mixed layer [280]. Following the same strategy, after electrografting of the sterically hindered 4-(tri-isopropylsilyl)ethynyl)phenyl groups, the cleavage of the isopropyl groups and further grafting of nitrophenyl groups also permit mixed aryl layers [246]. However, in the latter two investigations, the possibility that the second diazonium attacks the first grafted groups in parallel with the surface was never considered.

As illustrated in this section, it is now possible to tailor the properties of the films obtained by reduction of diazonium salts to characterize the film thickness, the surface concentration, and the deposited mass and to control the presence of different groups in the film.

3.7 MECHANISMS INVOLVED IN THE GRAFTING

We now describe the different mechanisms that have been proposed for the grafting of diazonium salts on different materials.

3.7.1 PREADSORPTION OF THE DIAZONIUM

Adsorption of a diazonium cation has been described as the first step of the process. However, neither Raman nor photoluminescence spectra of graphene [68,281] can be taken as proof of preadsorption of the diazonium salt because experiments were performed in pure water, where, at pH 11, the species exist as the diazohydroxide and the diazoate in solution. In this case, a charge transfer complex is unlikely between graphene and these species. Anthraquinone-2-diazonium salt (in ACN, where the diazonium is the reactive species) has been adsorbed onto an edge plane pyrolytic graphite electrode to form a thin unreacted submonolayer film. After transfer to a buffer solution containing no diazonium salt, the adsorbed material was thermally decomposed to complete the grafting procedure [240]. In this case, the adsorption is most likely due to the anthraquinone group and should be duplicated to explore whether the same experiment would be successful with, for example, **DNO₂**. Submonolayers were obtained on amorphous carbon substrates from very low concentrations of **DNO₂**, and their formation was followed by EQCM. The Γ versus c(**DNO₂**) data could only be fitted with a preadsorption step followed by the reaction of the adsorbed species [62]. Finally, by investigating the corrosion of iron (in 10^{-2} M H_2SO_4), it was shown that 2,6-dimethylbenzenediazonium (**D(2,6-CH₃)₂**) cannot attach to surfaces due to the steric hindrance of the radical by the two methyl groups and therefore did not provide any protection against corrosion. This would have been the case if (**D(2,6-CH₃)₂**) had been adsorbed on the surface [41]. Therefore, evidence for the preadsorption of diazonium salts before the grafting step are rather scarce and further investigations are needed.

3.7.2 The Existence of a Diazenyl Radical

The one electron reduction of a diazonium salt Ar–N=N⁺ could lead to a diazenyl radical Ar–N=N· (Figure 3.41). Such species have already been observed by radiolysis [282–283] and by esr at low temperature [284–285]. However, the simulation of the voltammogram of **DNO₂** pointed to a concerted electrochemical mechanism without intermediacy of the aryldiazenyl radical. The difference between these experiments was assigned to the very different driving forces of the different reactions; the cleavage passes from concerted to stepwise as the driving force becomes larger and larger [45].

3.7.3 The Mechanism of Multilayer Formation

The formation of multilayers is presented in Figure 3.42 [126]. Reaction [R15] involves the transfer of 2e⁻ and the formation of two aryl radicals **1**, where one of them attaches to the surface **2**. These electrons can be obtained from an electrode, a reducing substrate or a reducing reagent in solution. At this point, there are two possible reaction pathways. In route A, the first grafted aryl group is attacked by a radical [R17] leading to a dimeric cyclohexadienyl radical **3** that can be reoxidized

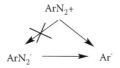

FIGURE 3.41 Stepwise versus concerted formation of an aryl radical from a diazonium salt.

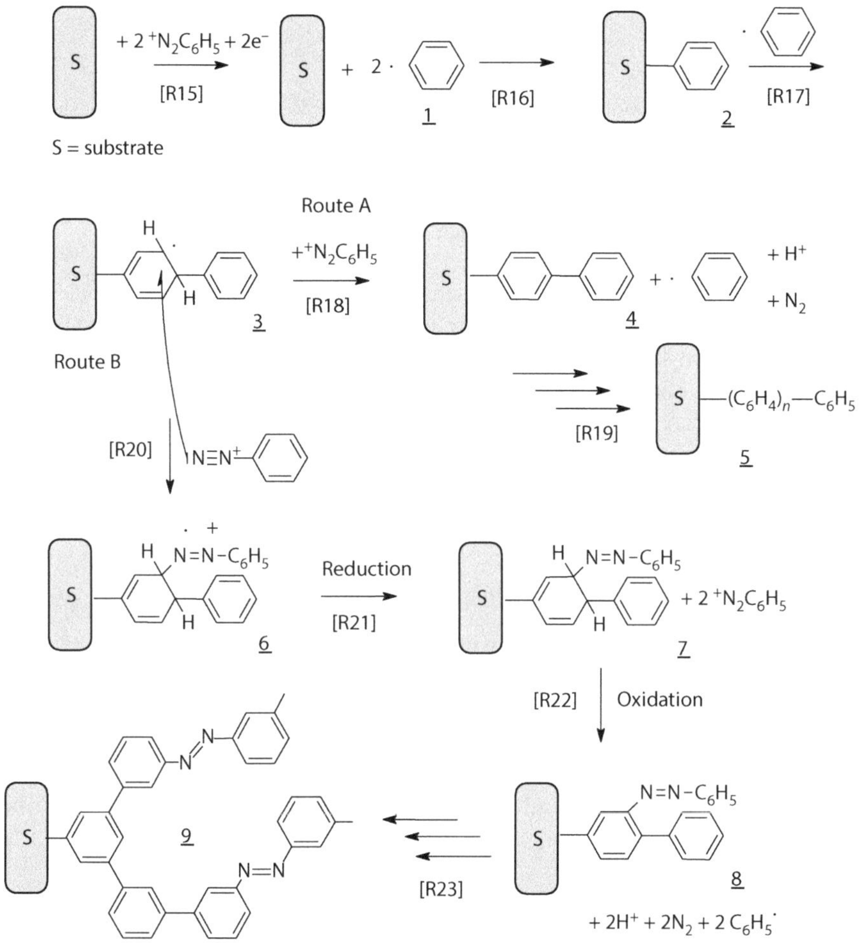

FIGURE 3.42 Schematic representation of the grafting reaction and the formation of multilayers.

[R18] by a diazonium. The products of route A are a biphenyl bonded to the surface **4**, an aryl radical that can start a chain reaction, nitrogen and a proton. In route B, the cyclohexadienyl dimer **3** is attacked by a diazonium cation [R20], leading to a radical cation **6** that will be reduced [R21] by the metal or carbon of the electrode connected or not or by electron exchange with an aryl radical. The cyclohexadienyl compound **7** can be reoxidized [R22] to an aromatic ring by reduction of another diazonium cation leading to **8** and again to an aryl radical. Due to the formation of aryl radicals in reaction [R18 + R22], the growth of the oligomers can continue along [R19 + R23]. This mechanism accounts for the presence of both biphenyl and azobenzene structures in the film and for the growth of the film under both electrochemical and spontaneous conditions due to the chain character (formation of the aryl radical in both route A and route B). This chain reaction is responsible for the

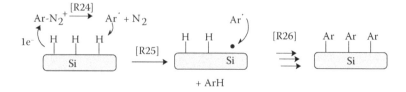

FIGURE 3.43 Electrografting of diazonium salt on *Si*-**H**.

continuing growth of the film even when the current is very small [286]. A similar chain mechanism has been shown to occur during the formation of biphenyl by reaction of diazonium salts with phenols [287].

The faradaic efficiencies of the reaction, measured by EQCM [179], are very high in ACN where up to 85% of the electrons are used in the grafting reaction and therefore only 15% in side reactions (e.g., dimerization, hydrogen atom abstraction from the solvent, etc.). This efficiency decreases dramatically in aqueous solution to 0.07%.

3.7.4 THE MECHANISM ON SI–H

The mechanism on Si–H is somewhat different, as shown in Figure 3.43 [100]. Electron transfer to the diazonium cation produces an aryl radical [R24]. This radical abstracts a hydrogen atom from the *Si*–**H** surface [R25] to give a silyl radical. Coupling [R26] of an aryl radical with a silyl radical completes the grafting of the aryl group. It is likely that the same mechanism accounts for the grafting on PMMA [90]. The growth of the film follows the reactions already described (Figure 3.42).

3.7.5 THE MECHANISM ON AU

During the spontaneous grafting on Au, the metal cannot be the reducing agent ($E°(Au^{3+}/Au) = 1.5$ V). A first mechanism postulated that Au only acts as a conductor and that adventitious impurities are the reducing agents [257]. This mechanism was supported by the observation of the open-circuit potential (OCP), where a positive shift was observed after the addition of the diazonium salt followed by a slow decrease. This mechanism was later challenged through careful investigation in ACN and aqueous solution [112]. This study proposed both the reaction of aryl cations with gold (heterolytic dediazonation) and the direct reaction of the diazonium salt with Au to give *Au*–N=N–$C_6H_4NO_2$ (Figure 3.16c).

3.7.6 THE MECHANISM ON HIGHLY ORIENTED PYROLYTIC GRAPHITE AND GRAPHENE

A radical mechanism is also involved on Highly Oriented Pyrolytic Graphite (HOPG) and graphene. However, as explained in Section 3.5, the reactivity is higher on the edge, and on the monolayer plane than on the bilayer plane [129].

As shown in this section, the mechanism of multilayer formation appears to be well established except on gold where results are not in agreement.

3.8 POLYMER GRAFTING TO DIAZONIUM-MODIFIED ELECTRODES

Among the many applications of diazonium salts, the attachment of polymers to surfaces seems to be one of the most interesting and has been widely investigated. This section is devoted to this topic.

3.8.1 Introduction to Polymer Grafting and Scope

Due to the bifunctionality of substituted aryl diazonium cations, they have been employed for polymer grafting from the early years of their modern surface chemistry when an epoxy resin was grafted to 4-aminophenyl-modified carbon fibers [288]. Because the dediazonation provides reactive radical species toward surfaces, one can take advantage of the functional group in the para position to attach polymers and, therefore, design new materials [6,118,289].

The methods for attaching polymers to surfaces fall into two categories: the so-called *grafting onto* and *grafting from* routes (Figure 3.44). In the former approach, preformed polymers react with aryl-modified surfaces having terminal reactive groups (toward the polymer). This concerns both the use of diazotized surfaces or diazotized polymers (see Section 3.8.2). In the latter, diazonium cations equipped with polymerization initiators are employed to modify surfaces, which in turn act as macro-initiators. With either method (grafting from and grafting onto), many systems have been described in the literature and reviewed [6,118,289]. In the following, we briefly summarize the concepts for polymer grafting through aryl layers and concentrate on case studies relevant to electrochemistry and electroanalysis applications.

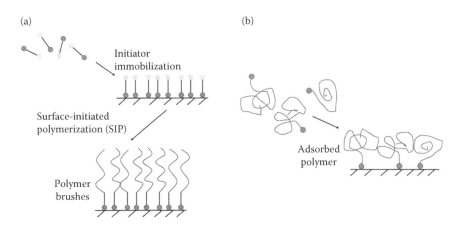

(a)   (b)

Initiator immobilization

Surface-initiated polymerization (SIP)

Adsorbed polymer

Polymer brushes

FIGURE 3.44 General methods of attaching polymers to surfaces: (a) grafting from and (b) grafting onto.

3.8.2 METHODS OF GRAFTING PRE-FABRICATED POLYMERS TO SURFACES (GRAFTING ONTO APPROACH)

Preformed polymers can be tethered to surfaces through aryl layers via a range of strategies. The first reported method concerned epoxy attachment to aminophenyl-modified carbon fibers $CF–NH_2$ [288]. Grafted nitrophenyl groups $CF–NO_2$ were reduced electrochemically to an amine therefore providing surface-aminated carbon fibers $CF–NH_2$ that readily reacted with epoxy resins. The method has just been simplified using p-phenylenediamine ($NH_2–C_6H_4–NH_2$) which was diazotized *in situ* with isoamyl nitrite and the resulting diazonium cation ($^+N_2–C_6H_4–NH_2$) reacted with carbon fibers "on water" [290].

Photografting polymers requires a surface-tethered photosensitizer. This can be achieved with diazotized benzophenone for the grafting of PS to iron [291] or PEG to SS [118].

Polymers can be grafted by click chemistry, an efficient strategy for making nanocomposites using diazonium-modified substrates [292–295]. $Au–N_3^+$ groups were clicked to ethyleneglycols using mono and bispropargyl functionalized oligo (ethylene glycol), OEG, and perfluorinated ethylene glycol (FEG) [295]. OEG and FEG grafts gave hydrophilic (static water contact angle $\theta = 48.7°$) and relatively hydrophobic ($\theta = 83.0°$) surfaces.

For carbon nanotubes, organosoluble PS-SWCNT nanocomposites [292] have been prepared that could be efficiently transformed to highly hydrosoluble, pH-responsive materials via direct sulfonation of the grafted PS chains [296]. In another approach, $MWCNT–C≡CH$ was clicked with azide-functionalized poly(glycerol methacrylate) and the final *hairy* MWCNTs were used for the *in situ* synthesis and immobilization of monodisperse 3 nm-sized palladium nanoparticles. The final heterostructure had a high catalytic activity (77%) for the C–C coupling Suzuki reaction between 4-bromobenzene and phenyl boronic acid [293].

Another emerging approach depends on the interfacial reaction of diazotized materials. The synthesis of *PEGylated* nitrobenzene ($PEG–C_6H_4–NO_2$) has been described. It was further reduced to obtain $PEG–C_6H_4–NH_2$ prior to diazotization by $NOBF_4$ to obtain $PEG–C_6H_4–N_2^+$ (**DPEG**) [224]. The derivatization of small diameter carbon nanotubes via electrochemical reduction of the **DPEG** was as high as 1 of 20 carbon atoms in the NTCs.

Diazotized dendrimers were reduced with Au(III) [297] or Pd(II) [298] to provide the corresponding *Au(np)* or *Pd(np)* core dendrimers. Palladium-grafted dendrimer nanoparticles catalyze the reduction of C=C and C≡C at room temperature. The Pd-core dendrimer was also found to be a good catalyst in Suzuki, Stille, and Hiyama coupling reactions [299] with a good recovery and recyclability in Suzuki coupling reactions.

Recently, layer-by-layer assemblies of polystyrene-based diazonium salt (**DPS**) and sulfonated reduced graphene oxide (SRGrO) were deposited on quartz, silicon, and ITO substrates [300]. The **DPS** was prepared from polystyrene (PS) through nitration, reduction, and diazotization reaction, whereas SRGrO was pre-reduced with $NaBH_4$, modified with diazonium salt of sulfanilic acid, and reduced by hydrazine. Besides a spontaneous reaction between **DPS** and SRGrO, the

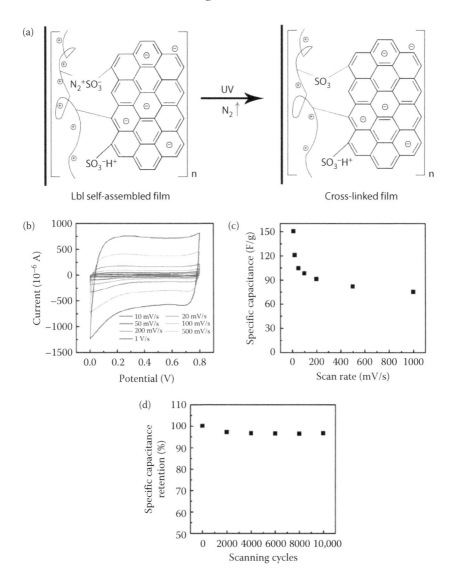

FIGURE 3.45 Preparation and performance of photo-cross-linked layered PS-graphene films. (a) Schematic illustration of the preparation procedure, (b) CV curves, (c) specific capacitance of the multilayer film with nine bilayers under different scanning rates, (d) cyclic test of the multilayer film with nine bilayers at the scan rate of 1000 mV/s. (Adapted from Xiong, Z., T. Gu, and X. Wang. *Langmuir* 30, 2014: 522–532.)

self-cross-linking of PSDAS, the authors brought strong supporting evidence for a UV-assisted reaction between the diazonium from **DPS** and the sulfonate from SRGO (Figure 3.45a):

$$PS\text{-}N_2^+ + {}^-O\text{-}SO_2\text{-}GrO \rightarrow PS\text{-}O\text{-}SO_2\text{-}GrO$$

These multilayered films were coated on ITO electrodes and characterized by cyclic voltammetry. The resulting voltammograms (scan rate in the 10–1000 mV/s range) displayed the nearly rectangular shape of Figure 3.45b, characteristic of electrochemical double-layer capacitors. The specific capacitance of cross-linked films with nine bilayers was 150.4 F/g in 1.0 M Na_2SO_4 (scan rate of 10 mV/s, Figure 3.45c). The cross-linked layered material exhibited superior cyclic stability with a 97% capacitance retention even after 10,000 charge – discharge cycles (Figure 3.45d). The cross-linked network structure of self-assembled films and high double-layer capacitance of SRGO nanosheets make the multilayer films a promising material for supercapacitor electrodes.

3.8.3 METHODS OF SURFACE-CONFINED POLYMERIZATION

Diazonium-modified materials served as macroinitiators of atom transfer radical polymerization (ATRP) (see Section 3.8.3.1), reversible addition-fragmentation chain transfer (RAFT) [301–302], radical photopolymerization (see Section 3.8.3.2), initiation, transfer, and termination (INIFERTER) [303,304], anionic polymerization [305], electropolymerization [306–308], and oxidative polymerization of conjugated monomers [309].

3.8.3.1 Atom Transfer Radical Polymerization

3.8.3.1.1 Method

ATRP has emerged as one of the most important and investigated methods of controlled radical polymerization since its discovery by the Sawamoto [310] and Matyjaszewski teams [311]. It was applied to surfaces only a few years only after its take off [312]. The reader is referred to References [313–314] that review important advances in surface-confined ATRP.

As far as diazonium surface chemistry is concerned, electrografting ATRP initiator groups [315] by electroreduction of aryl diazonium cations were first reported at the same time and independently from *in situ* diazonium formation and reaction with carbon black [316] to provide macro-initiators. Figure 3.46 displays a general mechanism of ATRP initiated by surface bound aryl groups from diazonium cation precursors. The final halogenated group can be borne by the diazonium cation [315] or a grafted hydroxyethyl phenyl group and can be reacted with 2-bromoisobutyryl bromide to provide a bromine-containing ester [317]. It was demonstrated that depending on the surface concentration of the initiator, the attached polymer chains could be in the mushroom or brush conformation [317]. The control over the ATRP process and linear growth of film thickness versus time was investigated for PBMA chains grafted on iron [318].

Atom transfer radical polymerization initiated at aryl-modified surfaces opened up many avenues in materials science in general and electrochemistry in particular. The following materials have been grafted with reactive and functional polymers via ATRP: glassy carbon [319–322], carbon fibers [187,323], carbon black [316], vertically aligned [324] and dispersed [325] carbon nanotubes, graphene [326], gold [327–328], iron [315,318–319], SS plates [329] and stents [330], nitrogen-doped ultrananocrystalline diamond [331], and nanodiamond particles [332]. In all of these

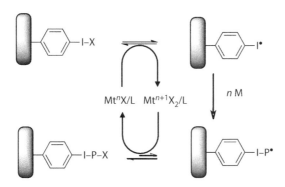

FIGURE 3.46 Idealized mechanisms of vinylic polymer grafting via ATRP initiated by diazonium-based initiators. I-X: initiator group; X: halogen; Mt: metal; L: ligand; M: monomer; P: polymer graft. (Adapted from Gam-Derouich, S. et al. *Aryl Diazonium Salts: New Coupling Agents in Polymer and Surface Science*, Chehimi, M. M. (Ed.), 2012, Chap. 6, pp. 125–157. Copyright Wiley-VCH Verlag GmbH & Co. KGaA. Reproduced with permission.)

publications, Cu(I) was used as the ATRP catalyst. Cu(I) can also be generated by electrochemical reduction of Cu(II) in the presence of aryl-modified surfaces. This elegant approach is termed *electrochemically mediated ATRP* (eATRP); it is detailed in Section 3.8.3.1.

3.8.3.1.2   Metal Chelatant Polymer-Grafted Fibers

Carbon fibers were electrografted with 1-phenylethylbromine layers for the ATRP of glycidyl methacrylate (GMA). The epoxy groups of PGMA-grafted chains were reacted with cyclam to provide chelatant, pH-responsive fibers (*CF*-**PGMA-Cy**) used as selective sensors for copper ions [187]. The *CF*-**PGMA-Cy** exhibited an uptake of copper of 28.6 mg/g at pH 5.2 (see XPS survey region in Figure 3.47 top right) and could be recovered up to five times without significant loss of activity. The *CF*-**PGMA-Cy** fibers with adsorbed copper ions were efficiently used as electrodes in electrochemical stripping voltammetry method (Figure 3.47, bottom right). Integration of the stripping peak indicated that the *CF*-**PGMA-Cy** system provides picomolar detection of copper ions.

In a similar approach, PGMA has been *grafted from* CF by eATRP as schematically shown in Figure 3.48 [323]. The epoxy groups were post functionalized by iminodiacetic acid groups providing chelatant diacidic repeat units for the uptake of nickel. The *CF*-**PGMA** fibers with immobilized Ni(II) were used as a working electrode at a potential of -0.7 V for 30 s in 0.5 M H_2SO_4 to obtain nano-nickel-coated CFs. The 80-nm-sized Ni nanoparticles have average crystallite size as low as 4.9 ± 0.4 nm. The Ni(II) adsorption fit a pseudo-second order kinetics mechanism and the maximum adsorption was found to be 53.3 mg/g (0.91 mM/g).

3.8.3.1.3   Redox Polymer Brushes

The PGMA brushes were grown by a SI-ATRP process on GC and post-functionalized in three steps in order to anchor the redox ferrocenyl and nitrophenyl groups.

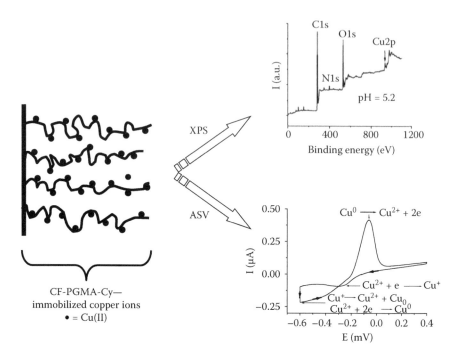

FIGURE 3.47 Schematic representation of *CF*-**PGMA-Cy**—immobilized copper complex and the corresponding XPS surface analysis and characterization by adsorption stripping voltammetry (ASV). XPS survey spectrum shows uptake of Cu2p from uptaken Cu(II). ASV (scan rate 10 mV s^{-1}) was conducted on a single *CF*-**PGMA-Cy-Cu** at pH 5. Cu(II) first reduces to Cu(I) which disproportionates to Cu(0) and Cu(II); Cu(II) is then reduced at −0.6 V (vs. Ag/AgCl). Electrochemical reduction to Cu(0) is responsible for the increase of the cathodic current. The stripping of Cu(0) to Cu(II) is observed at −0.06 V. (Adapted from Mahouche Chergui, S. et al. *Carbon* 48, 2010: 2106–2111.)

The multistep strategy for designing redox PGMA brushes is displayed in Figure 3.49 together with the electrochemical activity of the bi-functionalized brushes. The rationale for selecting PGMA is that it represents a reactive platform by ring opening of the epoxide group. This group is known to readily react with nucleophiles such as carboxylates, alcohols, amines, thiols, and azides. The reaction of the latter group offers an azide-functionalized polyglycerol which can further react via its OH pendant group. With this strategy in hand, it was possible to attach two redox groups to the repeat units of former PGMA brushes [321].

3.8.3.2 Surface-Initiated Radical Photopolymerization

3.8.3.2.1 Method

Surface-initiated radical photopolymerization (SIPP) is a simple and extremely efficient method to attach reactive and functional polymers to surfaces. The growth of PS, PMMA, and PHEMA was reported at a gold surface grafted with the 4-benzoylphenyl groups from the parent diazotized benzophenone for the first time in

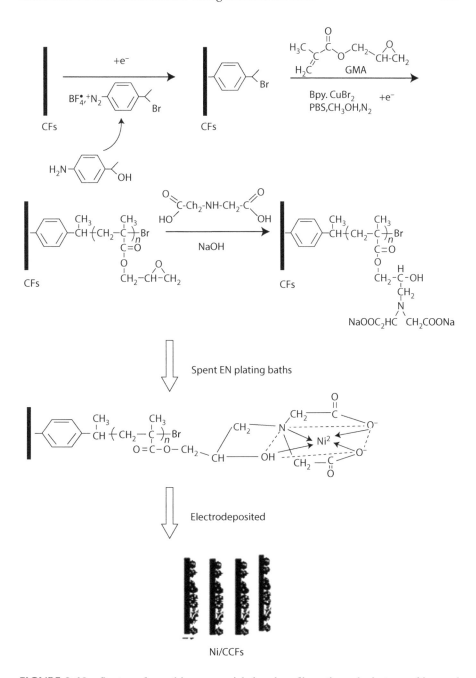

FIGURE 3.48 Strategy for making nano-nickel carbon fibers through electrografting aryl groups followed by eATRP of glycidyl methacrylate. (Adapted from Jin, G.-P. et al. *J. Appl. Electrochem.* 44, 2014: 621–629.)

(A)

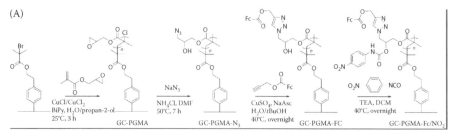

(B)

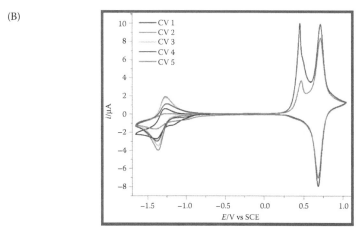

FIGURE 3.49 (A): poly(glycidyl methacrylate) brushes grown by atom transfer radical polymerization from initiator-modified glassy carbon substrates followed by post-polymerization modifications to create dual-functional polymer brushes. (B): First (CV1), second (CV2), third (CV3), fourth (CV4), and fifth (CV5) cyclic voltammograms of **GC-PGMA-Fc/NO2** recorded at a sweep rate = 0.02 V s^{-1} in 0.01 M Me$_4$NBF$_4$/MeCN. The first three cycles are restricted to the reductive region of the nitrobenzene group (-0.65 V $\rightarrow$ -1.65 V $\rightarrow$ -0.65 V), while the last two cycles include the positive potential region of the Fc group (-0.25 V $\rightarrow$ -1.65 V $\rightarrow$ 1.05 V $\rightarrow$ -0.25 V). (Adapted from Lillethorup, M. et al. *Macromolecules* 47, 2014: 5081–5088.)

2010 [333]. *N,N*-dimethyl aniline served as hydrogen donor in the SIPP process. The PHEMA grafts proved to be very resistant to non-specific protein adsorption. Many initiating systems were proposed and most of them used type II photoinitiators (e.g., an initiator that requests the use of a photosensitizer to strip a hydrogen). Among the various surface-initiating systems that have been developed [333–334] (see Section 3.8.3.2.2), one of the most efficient consists of the attachment of *N,N*-dimethylaminophenyl (DMA) groups by electroreduction of the diazonium precursor and use of free benzophenone as a photosensitizer in the polymerization medium [335]. The reason for this choice lies in the fact that the formation of semi-pinacol ((C$_6$H$_5$)$_2$–C·–OH) after stripping a hydrogen from the grafted DMA provides an efficient macro-initiator surface-DMA, whereas the solution semi-pinacol cannot initiate radical polymerization, which would provide free polymers in solution. It follows that polymerization is confined to the surface.

The strategy for making PHEMA grafts is displayed in Figure 3.50 together with surface analysis of a gold electrode grafted with PHEMA through a photoinitiator aryl layer. The efficiency of the system was investigated by changing the initial benzophenone to monomer molar ratio (in %).

3.8.3.2.2 Application of SIPP to the Design of Electrochemical Sensors Based on Molecularly Imprinted Polymer Grafts

Molecularly imprinted polymers (MIPs) have attracted a great deal of interest in recent years due to their ability to recognize specific and selective molecules, proteins, and even microorganisms, with excellent detection limits [336]. These sensing materials can be prepared as powders, colloids, and ultrathin films. The latter option is particularly interesting because it relies on diffusion of the analytes to the artificial receptor sites within the thin sensing layers and facilitates the making of nanostructured MIP grafts [337]. This diffusion process is known to be much faster compared to the diffusion within the bulk observed for powder samples. In addition, ultra-thin MIP layers permit the detection of analytes by electrochemistry in the picomolar or subpicomolar range [338]. However, the stability and robustness of the sensing layers

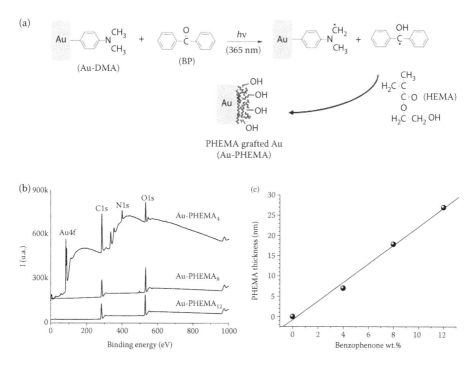

FIGURE 3.50 Upper panel (a): method of making polymer grafts by radical photopolymerization using a type II photoinitiating system. Lower panel: effect of benzophenone initial concentration on the surface composition of gold plates grafted with PHEMA; XPS survey regions are displayed in (b) and the deduced polymer thickness is plotted in (c). (Adapted from Gam-Derouich, S. et al. *Langmuir* 28, 2012: 8035–8045.)

depend on the efficiency of the process of tethering the layers to the electrodes. In this regard, time and effort were spent investigating aryl diazonium salts to initiate radical polymerization techniques at the surface of a variety of electrodes.

The general pathway to the MIP layers is schematically shown in Figure 3.51.

Many systems have been investigated. The first concerned the preparation of MIP grafts by ATRP against quercetin [340]. However, it soon became clear that SIPP was a much easier method of making such biomimetic grafts. Table 3.10 summarizes the MIP-grafted electrodes for the detection of molecules of biomedical and food importance. These systems are robust with a recovery up to five consecutive uses or more. They are highly selective toward the target molecules and can be employed in either PBS or complex matrices such as milk [341] or tea [342]. It is important to stress that, for example, the antioxidant molecule quercetin, a penta-hydroxylated flavonoid, is known to be a radical scavenger and inhibits the formation of a MIP graft by photopolymerization using the photo-initiating system grafted DMA/free benzophenone or the reverse system grafted benzophenone/free DMA. Instead, a diazotized dimethylethanolamine was found to act as a very efficient initiator in the presence of the hydrogen donor isopropyl-thioxanthone [342].

It is interesting to note that several parameters have an effect on the sensitivity of the MIP grafts. These include the template/functional monomer/cross-linker initial molar ratio [343], the pH of the rebinding aqueous medium or the nature of the organic solvent in which the rebinding of the template takes place, and the presence of nitrates in the detection medium [344]. The selective and highly sensitive detection of melamine by its corresponding MIP grafted to gold is shown in Figure 3.52a and highlights its selectivity over other molecules of similar shape and chemical structure. The system is highly selective and only melamine could be rebound by the MIP (Figure 3.52b).

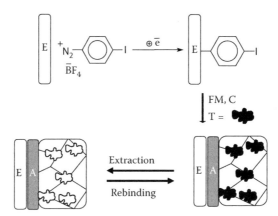

FIGURE 3.51 Strategy for making MIP grafts on electrodes (E). I: initiator, FM: functional monomer, C: cross-linker, T: template molecule, A: aryl layer. (Adapted from Bakas, I. et al. *Surf. Interface Anal.* 46, 2014: 1014–1020.)

TABLE 3.10
Modification of Electrodes by Molecular Polymer Grafts by Photopolymerization (hν = 365 nm) Initiated by Aryl Layers

Surface	Coupling Agent (Method of Attachment)	MIP Formulation: T/FM/CM	Conditions: Solvent, Coinitiator, Temperature, Time	Application [References]
Glassy carbon	$^+N_2$–C$_6$H$_4$–C(=O)–C$_6$H$_5$ BF_4^- (CA, 5 min, Acetonitrile/ NBu_4BF_4)	Dopamine/ MAA/ EGDMA (1/4/15)	Methanol CHCl$_3$ (80/20), DMA, RT, 1000 s	Detection in 0.1 M PBS [339]
Au	N≡N$^+$–C$_6$H$_4$–N< BF_4^- (CV, 30 cycles; CA, 5 min)	Quercetin/ MAA/ EGDMA (1/4/15)	Methanol/ CHCl$_3$ (70/30), ITX, RT, 1000 s	Detection in 0.1 M PBS [339]
	N≡N$^+$–C$_6$H$_4$–N< BF_4^-	Melamine/ MAA/ EGDMA (1/3/5)	DMSO, benzophenone, RT, 800 s	Solvent effect: electroanalysis [344]
	$^+N_2$–C$_6$H$_4$–C(=O)–C$_6$H$_5$ BF_4^- (CA, 5 min)	Melamine/ MAA/ EGDMA (1/3.2/5)	Methanol/ CHCl$_3$ (70/30), DMA, RT, 800 s	Detection in PBS and milk [341]
	$^+N_2$–C$_6$H$_4$–C(=O)–C$_6$H$_5$ BF_4^- (CA, 5 min)	Dopamine/ MAA/ EGDMA (1/4/15)	Methanol/ CHCl$_3$ (80/20), DMA, RT, 1000 s	Detection in PBS [343]
	$^+N_2$–C$_6$H$_4$–C(=O)–C$_6$H$_5$ BF_4^- (CA, 5 min)	Gallic acid/ MAA/ EGDMA (1/4/20)	THF, DMA, RT, 1100 s	Detection in PBS [339]
Au coated with Au NPs	$^+N_2$–C$_6$H$_4$–C(=O)–C$_6$H$_5$ BF_4^- (CA, 5 min)	Dopamine/ MAA/ EGDMA (1/4/15)	Methanol/ CHCl$_3$, DMA, RT, 1000 s	Detection in PBS [345]

(Continued)

TABLE 3.10 (*Continued*)
Modification of Electrodes by Molecular Polymer Grafts by Photopolymerization (hν = 365 nm) Initiated by Aryl Layers

Surface	Coupling Agent (Method of Attachment)	MIP Formulation: T/FM/CM	Conditions: Solvent, Coinitiator, Temperature, Time	Application [References]
ITO	(CA, 5 min)	Dopamine/ MAA/ EGDMA (2/8/30)	Methanol/ CHCl₃ (80/20), DMA, RT, 1000 s	Detection in 0.1 M PBS [339]
		Quercetin/ MAA/ EGDMA (1/4/15)	THF, ITX, RT, 2000 s	Detection in PBS, tea [342]

Source: Reproduced from Bakas I. et al. Molecularly imprinted polymeric sensings layers grafted from aryl diazonium-modified surfaces for electroanalytical applications. A mini review. *Surf. Interface Anal.* 46, 2014: 1014–1020, with permission.

CA: Chronoamperometry; CV: cyclic voltammetry; DMA: *N,N'*-dimethylaniline; DMSO: dimethylsulfoxide; EGDMA: ethylene glycol dimethacrylate; ITX: isopropyl thioxanthone; MAA: methacrylic acid; THF: tetrahydrofuran; VP: 4-vinylpyridine.

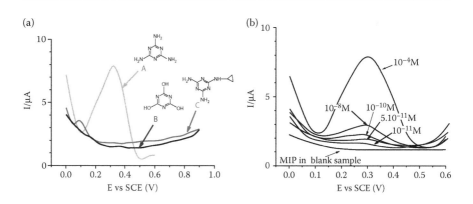

FIGURE 3.52 Performance of a MIP graft directed against melamine. (a) Selectivity: voltammograms of Au-MIPs after incubation in 10^{-4} mol L^{-1} solution of (A) melamine; (B) cyanuric acid; (C) cyromazine. (b) Square wave voltammograms (SWV) of Au-MIPs after incubation in methanol solutions of melamine with different concentrations in the 10^{-4}–10^{-11} mol L^{-1} range. SWV parameters: step potential 2 mV, pulse amplitude 25 mV, frequency 80 Hz. The oxidation peak intensity versus melamine concentration gave a linear semi-logarithmic correlation of I (μA) = 0.582 log (c) +21.948; r = 0.9936. The limit of detection was found to be 1.75 pmol/L. (Adapted from Bakas, I. et al. *Electroanalysis* 27, 2015: 429–439.)

3.8.3.3   Oxidative Polymerization of Conjugated Monomers

Grafting conjugated polymers to diazonium-modified surfaces was first described by electrografting of diphenylamine groups which served to tether the polyaniline (PANI) chains during their synthesis by electropolymerization [306]. Electrografting *in situ* generated a 2-amino *tert*-thiophenyldiazonium cation and resulted in an ultrathin oligothiophene graft onto glassy carbon, Au and Pt electrodes as discussed in Section 3.6.3 [270]. The formation of poly(3,4-ethylenedioxythiophene) (PEDOT) brushes attached to *GC*-thienyl modified surfaces has been described via surface-initiated electrochemical polymerization [308] where three different thienyldiazonium salts were compared. The structural conjugation of the resulting monolayer made it possible to initiate the electrochemical polymerization of EDOT from the modified surface. The polymeric structure formed on the thienyl-modified surface reveals a better redox peak separation as compared to the classical polymer film on a bare electrode that can be assigned to its more ordered and uniform structure. Poly(para-phenylene) (PPP) by oxidative electropolymerization to gold electrodes modified by aryl layers from benzenediazonium salt has been discussed in Section 3.6.3, Figure 3.26.

Perhaps one of the most technologically important developments involving electropolymerization of conjugated monomers on diazonium-modified electrodes has just been reported [121]. The authors electrografted aryl multilayers onto ITO-coated flexible poly(ethylene naphthalate) (PEN) substrates using 4-pyrrolylbenzenediazonium salt, which was generated *in situ* from a reaction between the 4-(1H-pyrrol-1-yl) aniline precursor and sodium nitrite in an acidic medium [121]. The first aryl layer is attached to the ITO surface through In−O−C and Sn−O−C bonds, which facilitate the formation of a uniform aryl multilayer ~8 nm thick. The aryl-modified flexible ITO was further modified through attachment of the polypyrrole-silver nanocomposite film formed by photopolymerization of pyrrole using an AgNO photosensitizer (Figure 3.53). The polypyrrole-silver film has excellent adhesion to the underlying ITO surface compared to the nanocomposite film deposited on untreated ITO. Additionally, the entire flexible ITO could be bent without any sign of failure of the conductive nanocomposite coating. These conductive polymer nanocomposite-grafted flexible ITO sheets open new avenues in flexible organic electronics, gas sensors, and biomedical applications.

New trends in the combination of conjugated polymers with diazonium interface chemistry also include the design of core/shell CNT/PANI nanocomposites that have a protuberance-free PANI shell due to the attachment of film forming PANI during synthesis onto the diphenylamine (DPA) attached layer to CNTs [309]. The nanocomposites could be used to develop ultrasensitive voltammetric sensors [346].

The efficiency of DPA in anchoring the forming PANI chains layers has been explored in view of designing exfoliated clay-PANI nanocomposites [347]. In this work, instead of *in situ* generated diazonium, the diazotized DPA was first synthesized and intercalated with the bentonite clay by a cation exchange mechanism with sodium from the layered silicate. Such an intercalation increased the lamellar spacing between the bentonite lamellae and facilitated the *in situ* oxidative polymerization of aniline which resulted in exfoliation of the clay. The nanocomposite prepared

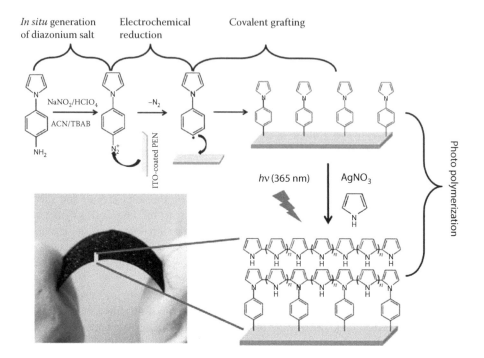

FIGURE 3.53 Schematic of *in situ* diazonium generation and electroreduction onto an ITO-coated flexible PEN substrate. The digital photograph shows the flexible ITO substrate grafted with silver-polymer nanocomposite film under bending stress. (Adapted from Samanta, S. et al. *Langmuir* 30, 2014: 9397–9406.)

with diazonium-modified clay exhibited a conductivity of 2×10^{-3} S/cm, which is five orders of magnitude higher than the corresponding nanocomposites prepared without any diazonium pretreatment of the clay. The nanocomposites could serve as conductive fillers or electrode materials.

3.8.3.4   Diazonium and Vinylics: the GraftFast Process

The grafting of vinylic (such as acrylonitrile, methylmethacrylate) compounds was the first electrografting reaction clearly characterized and identified as such [348]; it gives polymeric layers [4]. The anionic polymerization mechanism of this reaction implies oxygen- and water-free conditions (i.e., working in a glove box) [349]. This is a major drawback in industrial applications. However, when diazonium and vinylics are electrografted together modified surfaces can be obtained under ambient conditions. For example, a mixture of **DNO$_2$** and a vinylic compound (acrylonitrile, acrylic acid, hydroxyethylmethacrylate) in aqueous solution has been reduced by iron powder, ascorbic acid, or hypophosphorous acid. A grafted polymeric layer (Figure 3.54) including nitrophenyl groups has been characterized (patented as GraftFast™ process). The GraftFast process has been applied to many surfaces (metallic, carbon, semi-conductor, ceramic, or polymer) [79]. It can also be performed in an aqueous emulsion using cationic, anionic, or neutral

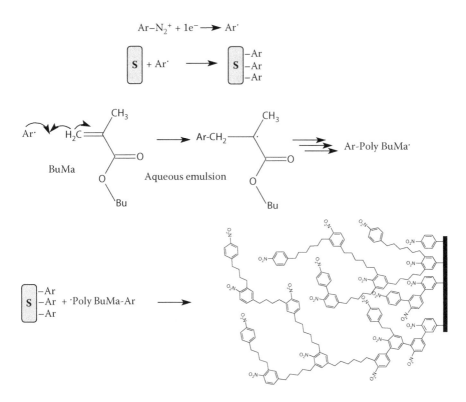

FIGURE 3.54   Proposed structure and mechanism of the Graftfast™ PHEMA films.

surfactants [350]. The mechanism has been carefully investigated [351]. Aryl radicals generated by reduction of diazonium salts bind to the surface and simultaneously initiate polymerization in solution. The resulting polymer chain radicals bind to the aryl-modified surface. Therefore, copolymers are prepared consisting of oligomer sequences bonded to the surface by aryl groups from the diazonium (Figure 3.54). The formation of the polymer shifts from an anionic polymerization in absence of diazonium to a radical polymerization that can be operated under ambient conditions.

This process of surface modification has been applied in particular to CNTs [352] for patterning metallic substrates [353] and for the removal of heavy ions from wastewater [354]. The GraftFast™ process was used to graft poly(acrylic acid) (PAA) on gold [355] and carbon felt [356]. The PAA grafts permitted efficient uptake of copper ions which were then released by electrolysis. Cu2p peak doublets were detected by XPS demonstrating the uptake of copper ions, and their absence after electrolysis due to the generation of hydrogen ions that switch PAA grafts from the carboxylate to the acidic form. The advantages of this "grafting from" method is the strong bonding to the surface as is seen in SIATRP (surface-initiated atom transfer radical polymerization), but is much faster and does not necessitate anhydrous conditions.

3.8.4 Conclusion

Silanes and many other compounds continue to serve successfully as coupling agents. From achievements over the past decade, it is clear that diazonium salts can also be regarded as new generation coupling agents for polymers due to the stable-layered materials they provide. The availability of many aromatic amines and ease of synthesis of diazonium salts make these compounds interesting due to the excellent adhesion they ensure for polymers to diazonium-modified surfaces. Thus, we anticipate several new electrochemical applications for polymer nanofilm-aryl primer conjugates.

3.9 OTHER METHODS AND REAGENTS FOR SURFACE MODIFICATION

In addition to diazonium salts, there are many other species that are able, once electrochemically activated, to modify the surface with aromatics, unsaturated and saturated groups. These surface modifications can be achieved by: (i) oxidation (amines, hydrazines, alcohols, carboxylates, carbamates, carbanions, Grignard reagents); (ii) reduction, some being closely related to diazonium salts (iodoniums, sulfonium) and other different species (vinylics, alkyl halide); (iii) spontaneously, chemically or photochemically (alcenes, azides, peroxides). Note that electrochemical oxidation methods can be applied only to the materials that can withstand quite positive potentials such as carbon or Pt. This section will describe these modifications [357].

3.9.1 Electrografting by Oxidation

3.9.1.1 Amines

Electrochemical oxidation of primary aliphatic and aromatic amines and their grafting onto material surfaces [288] has been reviewed [357]. Amino compounds with the general formula RNH_2, with R^- representing an aliphatic or aromatic group in millimolar concentrations in organic solvents are oxidized at potentials >1.0 V/SCE on a limited number of surfaces to a radical cation. After a rearrangement and the release of one proton, the aminyl radical attaches to the electrode surface (Figure 3.55). The reaction is also possible with aniline derivatives. Spontaneous grafting of amines is also possible, in particular on oxidized metallic surfaces though nucleophilic substitution and Michael additions on carbon [358–360].

Electrochemical grafting of amine compounds can be easily observed by cyclic voltammetry. Figure 3.56 shows the voltammograms obtained when a GC electrode is immersed in ACN + N-Boc-ethylene diamine solution [361]. During the first scan

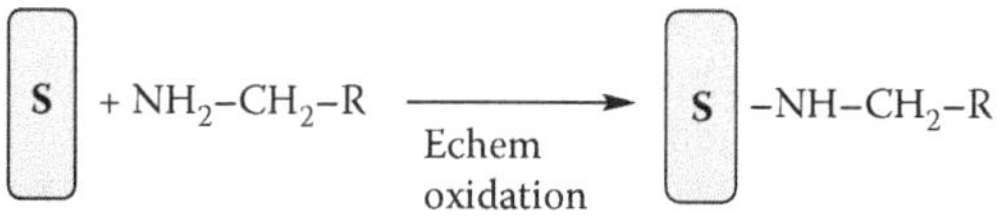

$$\boxed{S} + NH_2\text{--}CH_2\text{--}R \xrightarrow[\text{oxidation}]{\text{Echem}} \boxed{S}\text{--}NH\text{--}CH_2\text{--}R$$

FIGURE 3.55   Electrooxidative grafting of amines.

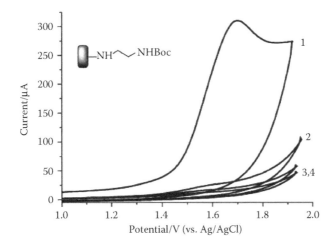

FIGURE 3.56 Cyclic voltammetry of a GC electrode in ACN of N-Boc-ethylenediamine solution. Cycles 1–4 are shown in the figure. (Adapted from Chrétien, J. M. et al. *Chem. Eur. J.* 14, 2008: 2548–2556.)

in the positive direction, an irreversible wave appears with the peak potential at ~1.7 V/Ag/AgCl. In the second and other consecutive scans, this wave diminishes drastically due to the passivation of the electrode surface with the organic film as in the case with aryl diazonium salts above.

Electrochemical oxidation of amines has been achieved on a number of surfaces including carbons, metals that withstand oxidation like platinum, as well as ITO and Si (Table 3.11). A number of different amines have been used; typical examples are reported in Table 3.12.

Different solvents have been used including aprotic solvents such as ACN, *N,N*-dimethyl formamide (DMF) [363], pure ethylenediamine [382,393], ethanol [364,386,394,395], and ionic liquids [396].

Spontaneous reactions involve a nucleophilic attack, for example, on the epoxy groups of GO [367,377–379] or a Michaël addition [359] on an activated double bond in carbon (Figure 3.57).

The electrochemical mechanism has been investigated [386] for amines with an α-methylene group, but not for aromatic amines. It involves the one electron formation of a radical cation that deprotonates to a carbon radical after tautomerization. The resulting aminyl radical binds to the surface (Figure 3.58). It was also shown that the reaction of secondary amines was more limited and that tertiary amines did not react [364].

An interesting mechanism is provided with ferrocifen (**1**, Figure 3.59) that is oxidized at +0.45 V/SCE, which is less negative than the aromatic amine. An aminyl radical is obtained through a proton abstraction and internal electron transfer from the ferrocene group to the amine; this radical reacts with the *Au* surface. The formation of multilayers was demonstrated by measuring the thickness up to 85 nm. The ferrocene group likely acts as an electron relay as in the case of thick layers formed

TABLE 3.11
Different Surfaces for Grafting Amines

Substrates	Comments	References
GC Spontaneous [358–359] Electrochemical [363–366]	Modified electrodes for analysis	[362–363,365–367]
	Attachment of Au-nP to carbon surfaces by electrografting	[368]
	Catalysis, electrocatalysis	[369–370]
Graphite powder	Electrografting, detection of cocaïne	[371]
Carbon felt	Electrografting, analysis	[372–373]
PPF	Modified electrodes Patterning of surfaces	[374–376]
Carbon fibers	Fibers for carbon–epoxy composites	[288]
Graphene oxide Spontaneous grafting	Selective dye adsorption	[367]
	Antibacterial membrane by nucleophilic substitution	[377]
	Nanocomposites by nucleophilic substitution of epoxide groups	[378–379]
CNT	Attachment of ethylenediamine for catalysis	[380]
	Attachment of ethylenediamine for analysis	[381]
Pt	Electrografting of pure ethylenediamine	[382]
ITO	Electrografting of dendrimers	[383–384]
Si	By electrografting	[71,385]

from diazonium salts. The formation of multilayers during the electrografting of amines has been demonstrated as expected from a radical mechanism [375].

Arylhydrazines can easily be electrochemically oxidized on a GC electrode [397–398] ($E_{pa} \sim + 0.2$ V/SCE for NH_2–NH–C_6H_4–NO_2 in aqueous solution); the surface is progressively blocked upon repetitive scanning. Modification of the surface was demonstrated by redox probe experiments and PMIRRAS. The surface concentration $\Gamma = 4.5 \times 10^{-10}$ mol cm^{-2} [399] was measured and thicknesses on the order of a monolayer were found. This process has been used for surface patterning by SECM [400]. The mechanism [401] that was established involves an aryl radical that is formed through an intermediate aryldiazene (Figure 3.60). Although an aryl radical is involved in the grafting process, monolayers are essentially obtained. This has been explained by the blocking of the electrode surface by a physisorbed layer consisting of organic solution products. In addition, the growth process is not able to induce a chain reaction as for aryldiazonium salts.

TABLE 3.12
Different Amines That Have Been Grafted on Surfaces

Amines	Comments	References
Butylamine	Comparison of the electrochemical behavior of primary, secondary, and tertiary amines	[386]
Ethylenediamine	Formation of a bonded polymer on the surface of the electrode	[382]
Diaminoalkanes (NH_2–CH_2–NH_2, $n = 7,10,12$)	Blocking properties of the film increase with the length of the chain	
Jeffamine (poly-oxypropylene diamine)	By spontaneous reaction with the carboxylic groups of GO	[387]
Amino acids	β-Alanine, aspartic acid, aminobenzoic acid	[366]
Ferrocifen (1-(4-aminophenyl)-1-phenyl-2 ferrocenylbut-1-ene)	At the oxidation potential of Fc, followed by internal electron transfer	[388–389]
Aromatic amines	-4-aminobenzoic acid	[14,390]
	-4-aminosulfonic acid	[391]
	-4-aminophosphonic acid	[392]

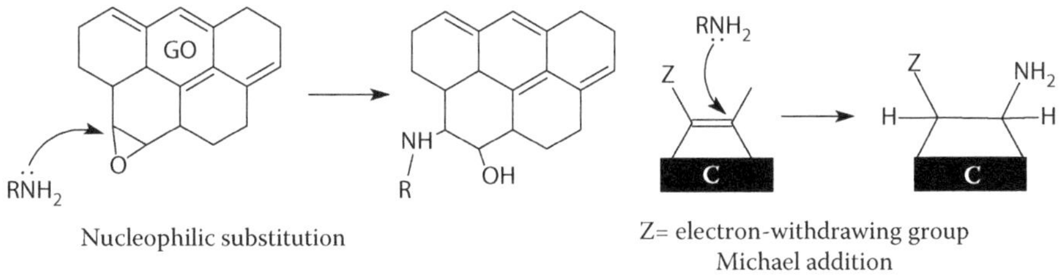

Nucleophilic substitution

Z= electron-withdrawing group
Michael addition

FIGURE 3.57 Spontaneous grafting of amines on carbon.

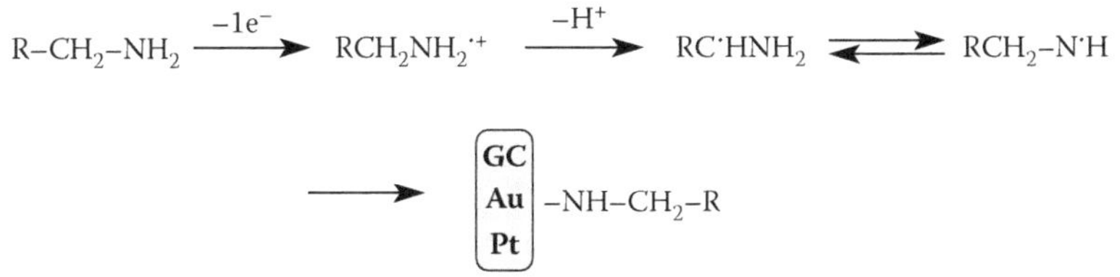

FIGURE 3.58 The oxidation mechanism and grafting of primary amines.

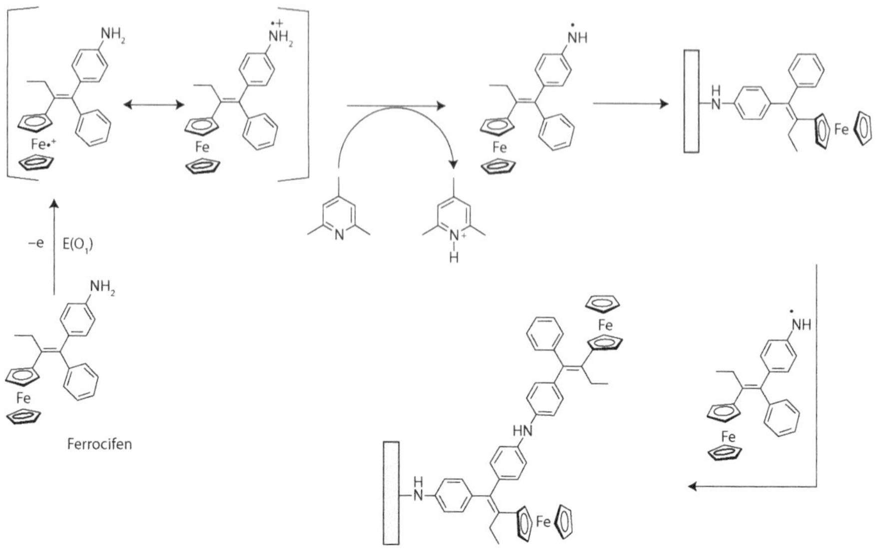

FIGURE 3.59   Electrografting of ferrocifen.

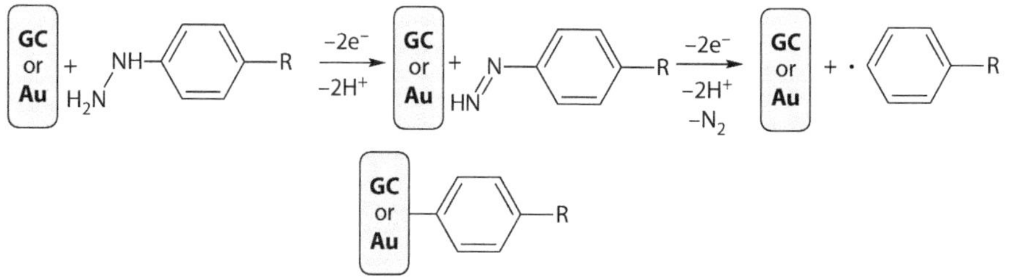

FIGURE 3.60   Electrografting of arylhydrazines.

3.9.1.2   Alcohols

Alcohols can be electrografted at very negative potentials [4,402] (Figure 3.61) and a small wave appears at the foot of the oxidation. The mechanism of this reaction has never been investigated, though it has been postulated that aromatic moieties at the surface of carbon are oxidized to radical cations that undergo a nucleophilic attack by alcohols [402]. This process has been applied to the grafting of diols and also to platinum dendrimer-encapsulated nanoparticles prepared within fourth-generation hydroxyl-terminated, poly(amidoamine) dendrimers [403]. These species are

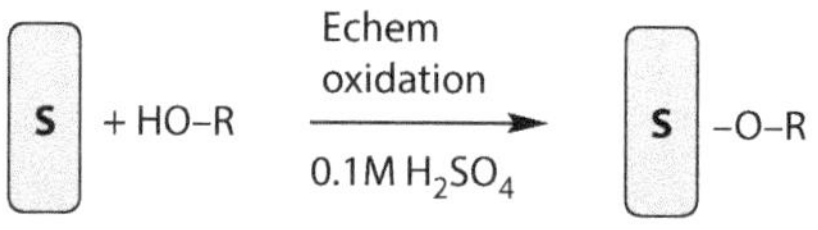

FIGURE 3.61   Electrografting of alcohols.

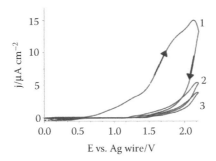

FIGURE 3.62 Grafting and cyclic voltammogram in neat 1,3 propanediol/ 0.1 M LiClO$_4$ on GC. (Adapted from Verma, P. et al. *Electrochim. Acta* 56, 2011: 3555–3561.)

electrocatalytically active for oxygen reduction. They remain strongly attached to the surface through 50 consecutive cyclic voltammograms as well as through sonication.

The GC electrodes have been electrografted by 1,3-propanediol (Figure 3.62) and further modified to give lithium alkyl carbonate–modified carbon with potential applications in batteries [404]. The modified surface has been characterized by XPS and redox probe investigations.

3.9.1.3 Carboxylates

The electrochemical oxidation of alkyl carboxylates is the well-known Kolbe reaction that leads to dimers. When the reaction is performed on a carbon electrode, grafting of benzyl [405–406] or alkyl [407] groups is observed (Figure 3.63). The voltammograms show an irreversible peak (~1 V/SCE) that decreases upon repetitive scanning, as seen with diazonium salts, and the grafted electrode blocks the electron transfer of Fe(CN)$_6^{3-}$, clearly indicating a modification of the electrode.

There have been debates as to whether carbocations R$^+$ or carbon radicals R· are responsible for the grafting reaction in understanding the mechanism. The important points of the discussion are:

a. Alkylcarboxylates give rise to products involving transient carbocations on oxidation, but do not attach to the GC surface.
b. It has been shown [408] that the formation of the radical is a stepwise reaction, with the RCOO· radical as an intermediate.
c. The decarboxylation rate constant of RCOO· radicals [409] has been measured by photochemistry and lies in the 10^9–10^{10} s^{-1} range. This implies that the radical is formed very close to the electrode and is prone to react with it.

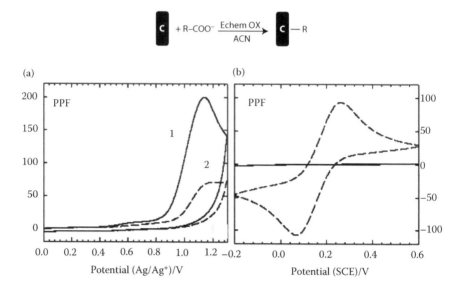

FIGURE 3.63 (a) First and second scan of naphthyl–CH_2–COO^- in ACN on a PPF electrode. (b) Modified **PPF–CH_2–naphthyl** electrode in an aqueous solution of $Fe(CN)_6^{3-}$/0.2 M KCl. Comparison at a bare PPF electrode (—). (Adapted from Brooksby, P. A., A. J. Downard, and S. S. C. Yu. *Langmuir* 21, 2005: 11304–11311.)

d. However, the radicals CH_3–$CH_2\cdot$ and $C_6H_5CH_2\cdot$are very easily oxidized (approx. +1.0 and +0.7 V/SCE for CH_3–$CH_2\cdot$ and C_6H_5–$CH_2\cdot$, respectively [48]), and could be oxidized to the carbocation $C_6H_5CH_2^+$. Therefore, a competition occurs at the electrode between the electron transfer and the reaction with the surface. An interesting investigation was performed to solve this problem [410–412] (Figure 3.64). It is, to our knowledge, the only electrocatalytic grafting in the literature.

Acetate (CH_3COO^-) ions [411] that are electroxidized to carbocations, as shown by the analysis of the reaction products, cannot be electrografted by direct oxidation of the acetate ion. However, the catalytic oxidation of acetate has been achieved by the ferrocene/ferricinium redox couple. The resulting voltammetric curves (a) are typical of redox catalysis, where the reversible ferrocene/ferricinium couple becomes irreversible and the anodic wave more than doubles in height as the concentration of acetate increases. In contrast, in redox catalysis without surface complication (b), the anodic peak of ferrocene decreases and finally disappears upon repetitive scanning (Figure 3.64) indicating the formation of an attached layer. This film was characterized by cyclic voltammetry (in the case of 4-nitrophenyl groups attached to the electrode) and by the inhibition of the wave of dioxygen (after electrografting of methyl groups by oxidation of acetate). Based on a thorough thermodynamic and kinetic investigation, the authors proposed that the formation of carbocations (responsible for the products in solution) and radicals (responsible for the grafting reaction) are in competition. The catalytic mechanism leading to the formation of radicals is shown in Figure 3.65.

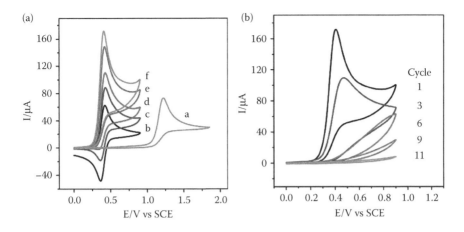

FIGURE 3.64 Cyclic voltammograms of tetrabutylammonium acetate on a GC electrode in ACN (a): (a) Direct oxidation. (b–f) Oxidation of 2 mM ferrocene at different concentrations of acetate: 0.57, 10.7, 23.1, 41.2 mM, respectively. (b) 2 mM ferrocene in the presence of 40 mM tetrabutylammonium acetate. Successive scans as indicated. (Adapted from Hernández-Muñoz, L. S. et al. *Electrochim. Acta* 136, 2014: 542–549.)

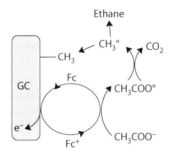

FIGURE 3.65 The electrooxidative grafting mechanism of acetate.

The process has been extended to unsaturated aliphatic carboxylates. For example, 4-pentenoic and 4-pentynoic acids have been attached to the GC surface and the presence of the unsaturated group was demonstrated by performing Heck and click reactions with reporter groups [413]. This reaction combined with the reduction of diazonium salt has been used to grow polymer brushes from surfaces under ATRP conditions, as shown in Figure 3.66. Starting from the diazonium salt of 4-aminophenylacetic acid, upon oxidative grafting of the carboxylic group, the phenyldiazonium groups is anchored to the surface. It can be further reduced to a radical that can start a Meerwein-ATRP reaction to form a film of methylmethacrylate ~60 nm thick [414]. Starting from the same grafted radical, it is possible to attach ferrocene or to perform a Sandmeyer reaction.

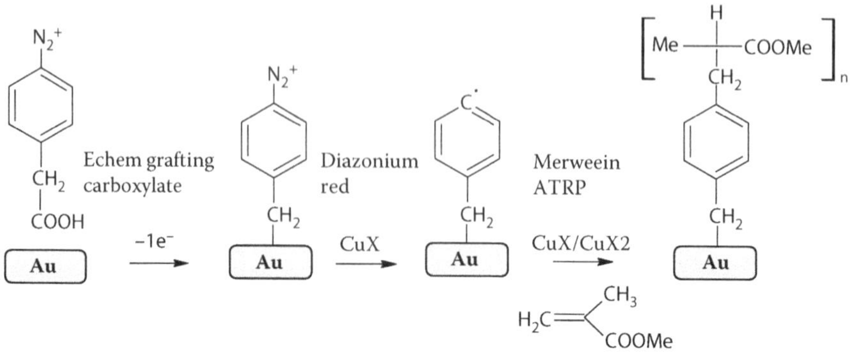

FIGURE 3.66 Meerwein ATRP from a grafted diazonium salt. (Adapted from Hazimeh, H. et al. *Chem. Mater.* 25, 2013: 605–612.)

Another reaction of diazonium carboxylate is presented in Figure 3.67. By heating the internal carboxylate diazonium salt, a diradical is obtained that reacts with carbon black. Transformation of the dimethoxybenzene group to *o*-quinone provides a modified carbon surface that can be used as a pseudocapacitance along the quinone/hydroquinone couple [415].

3.9.1.4 Carbamates

Carbamates (NH_2–COOH/NH_2–COO$^-$) can be electrografted on GC at approximately +1 V/SCE in aqueous solution to give secondary and tertiary amino groups on the surface, which have been characterized by XPS and by reaction with *o*-quinone. A mechanism has been proposed that involves the radical ·NH–COOH that reacts by hydrogen atom abstraction or attack on the quinonic structures of carbon [416]. The result of this reaction can be compared with the amination of GC performed by polishing in the presence of ammonia [417].

3.9.1.5 Carbanions

The carbanion of 5-nitroindole can be oxidized at E = +0.4 V/SCE in ACN and the voltammogram disappears after 15 cycles (Figure 3.68) [418]. The cleaned, modified electrode transferred to an ACN solution results in a voltammogram typical of the

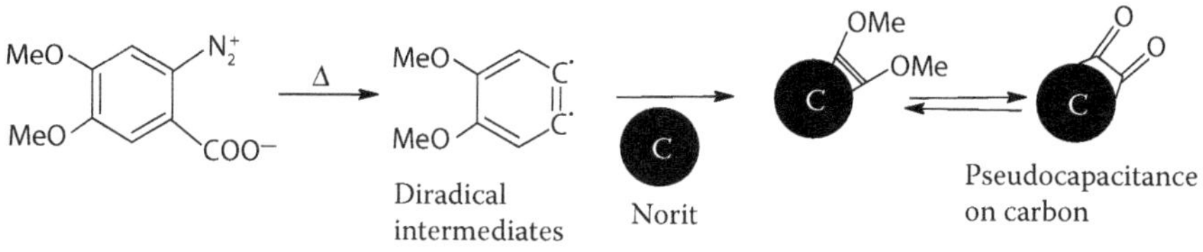

FIGURE 3.67 Modification of carbon black by heating a carboxylate diazonium. (Adapted from Lebègue, E. et al. *Electrochem. Comm.* 34, 2013: 14–17.)

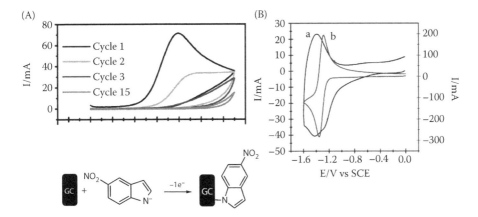

FIGURE 3.68 Electrografting of 5-nitroindole anion (NI⁻) on a GC electrode. (A) Repetitive cyclic voltammetry of NI⁻ in CAN. (B) (a) **GC-NI** electrode; (b) voltammogram of 1-methyl-5-nitroindole under the same conditions. (Adapted from Muñoz, L. S., C. Frontana, and F. J. González. *Electrochim. Acta* 138, 2014: 22–29.)

nitroindole moiety and the measured surface coverage $\Gamma = 4.8 \times 10^{-9}$ mol cm^{-2} corresponds to the formation of multilayers. Cyclodextrin has also been grafted from its carbanion [412].

In a similar way, upon oxidation the lithium stabilized carbanion of an ethynyl compound or the ethynyl compound itself ($E \sim 1.4$ V/SCE for Fc–C≡CH) leads to a radical that reacts with the surface of carbon as shown in Figure 3.69. The presence of the Fc group on the surface is ascertained by cyclic voltammetry and the surface coverage $\Gamma = 8.1 \times 10^{-10}$ mol cm^{-2} is equivalent to about 1.8 monolayers [419].

3.9.1.6 Grignard Reagents

Grignard reagents (aryl, vinyl, or alkyl) can be electrograted by oxidation on Si–H to give *Si*-alkyl, vinyl, or aryl multilayers through the formation of radicals [420–421]. The films are therefore quite similar to those obtained from diazonium salts. The drawback of the method lies in the necessity to work under oxygen- and water-free conditions to prevent the oxidation of Si and the hydrolysis of the reagent.

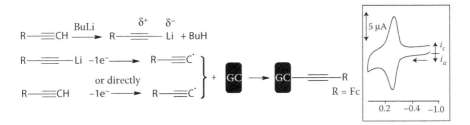

FIGURE 3.69 Electrografting of ethynyl anions, with the voltammogram of a GC–C≡C–Fc electrode in CH₂Cl₂. (Adapted from Sheridan, M. V. et al. *Angew. Chem. Int. Ed.* 52, 2013: 1–5.)

3.9.2 ELECTROGRAFTING BY REDUCTION

3.9.2.1 Ammonium, Phosphonium, Sulfonium, Iodonium, Xenonium, and Stibonium Salts

Closely related to diazonium salts [4], these salts differ by their reduction potential that ranges from +0.45 to −0.73 V/SCE for $(C_6F_5)Xe^+$, $(C_6F_5)–N_2^+$, $(C_6F_5)_2–I^+$, $(C_6F_5)_2–Br^+$, $(C_6F_5)_4–P^+$ [422]. Iodoniums have been investigated the most due to the following features:

a. They are reduced at more negative potentials than diazonium salts ($E_p = -0.37$ vs. +0.20 V/SCE for $(NO_2–C_6H_4)_2–I^+$ and $NO_2–C_6H_4N_2^+$, respectively, in ACN) [423,424].

b. They are also less reactive [425]; although they generate the same aryl radicals as diazonium salts. This difference probably stems from the partial reduction of the radical at the potential where they are formed.

c. At the different ArN_2^+ species, there is no spontaneous grafting on GC and Au which may be of interest for patterning surfaces.

The problem concerning $R–I^+–R'$ is which of R or R′ attaches predominantly to the surface, particularly when R = Ar and R′ = Alk. Starting from the aryl-alkyl-iododium salt of Figure 3.70 (R = −NO_2, R′ = −C≡C–(CH_2)_4–Cl), the surface concentrations are $\Gamma_{vol}^{Alk}=0.2$ and $\Gamma_{vol}^{Ar}=5\times10^{-10}\,mol\,cm^{-2}$ in agreement with the high reactivity of the nitrophenyl radical [424]. These results were supported by DFT calculations.

$(CF_3–C_6H_4)_2–I^+$ was grafted on graphene [426] at approximately −1.5 V/SCE to give $\Gamma_{surf} = 1.7 \times 10^{-10}$ mol cm^{-2}; this increases the work function of graphene by 0.5 V. The absence of a spontaneous reaction has been exploited for patterning ~400 μm lines on gold substrates by using SECM [427]. This is not possible with diazonium salts, where the entire surface is covered by spontaneous grafting.

3.9.2.2 N-Bromosuccinimide

N-bromosuccinimide can be reduced to a succinimidyl radical by reaction with Fc. This radical binds to Si–H to give a bilayer of Si-succinimide in a manner similar to diazonium salts [428].

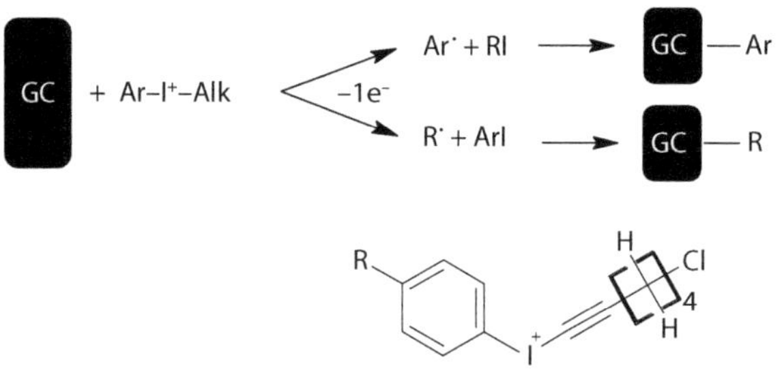

FIGURE 3.70 Grafting of iodonium salts.

3.9.2.3 Vinylics

Vinylics in the presence of diazonium salts were discussed in Section 3.8.

3.9.2.4 Alkyl Halides

Alkyl halides can be electrografted on metals [429] and carbon [430–431] at quite negative potentials. Examples include alkyl iodides (that are more easily reduced than bromo- or chloro- derivatives) with $E_p(IC_6H_{13}) = -2.5$ V/SCE ($n = 2$) and -2.3 V/SCE on Au ($n = 2$) and GC, respectively, but at more positive potentials (this is usually termed electrocatalysis) on Cu with $E_p(IC_6H_{13}) = -1.3$ ($n = 1$) [429], on Pd with E_p (IC_4H_9) = -1.8 V/SCE [432,433], and on Ag with E_p (IC_4H_9) = -1.1 V/SCE [434–435]. The number of electrons transferred at the peak potential is 2 on GC and Au, in agreement with the measured reduction potential of alkyl radicals (R·, $n = 1$). This potential can be approximated by that of the n-butyl radical (experimental) [436] in the $E° = -1.3/ -1.4$ V/SCE range and that of the ethyl radical (calculated) [48] with $E° = ~1.8$ V/SCE. At more negative potentials, the radical is reduced to the carbanion ($n = 2$) as in Figure 3.71.

This means that on Cu, the radical is not entirely reduced and can react on surfaces. However, grafting also occurs on gold in spite of the very negative reduction potential ~1 V more negative than the redox potential. In this case, the radical should react with the surface faster than being reduced. On carbon, the mechanism which is operative is the charging of the carbon followed by a heterogeneous SN_2-like attack on the alkyl iodide [437–439]. Using this method, alkyl iodides and also alkyl, benzyl, allyl, propagyl bromides [429–431,440] have been attached to carbon and metals. The multilayers that were formed are a typical feature of radical grafting reactions.

3.9.2.5 Indirect Electrografting with Diazonium Salts

As seen above, aryl radicals are very reactive species that readily bind to surfaces. They can also undergo crossover reactions to generate other radicals that can in turn bind to surfaces (Figure 3.2). Aryl radicals can abstract a hydrogen atom from the surface of Si–H to give a surface silyl radical that can in turn react with an alkyl/arylselenoether or a –C=C– bond to give a silicon-modified surface with, for example, an alkyl group as in Figure 3.72 and no aryl group. This is achieved in the presence of a large excess of the alkene in order to divert the reactivity of the aryl radical [441].

In another example, 2,6-dimethylbenzenediazonium tetrafluoroborate **D(2,6-Me)2** is reduced by electrochemistry in an ACN solution to the corresponding radical. This 2,6-dimethylphenyl radical cannot attach to surfaces due to the steric hindrance of the two methyl groups [178]. Instead, it abstracts a hydrogen atom from CH_3CN to give the cyanomethyl radical $·CH_2CN$ that reacts with the surface (Au, Cu, Si) as indicated in Figure 3.73 [24]. Further reaction of the $^-CH_2CN$ carbanion, obtained

$$RI + 1e^- \longrightarrow R· + 1e^- \longrightarrow R^-$$

FIGURE 3.71 Reduction of alkyl iodides.

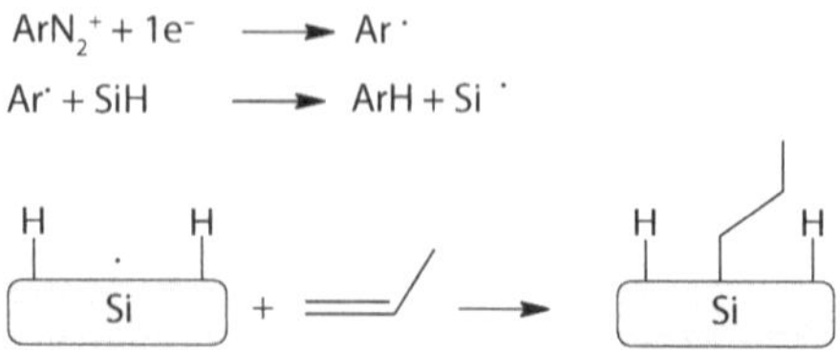

FIGURE 3.72   Indirect electrografting of alkenes on Si.

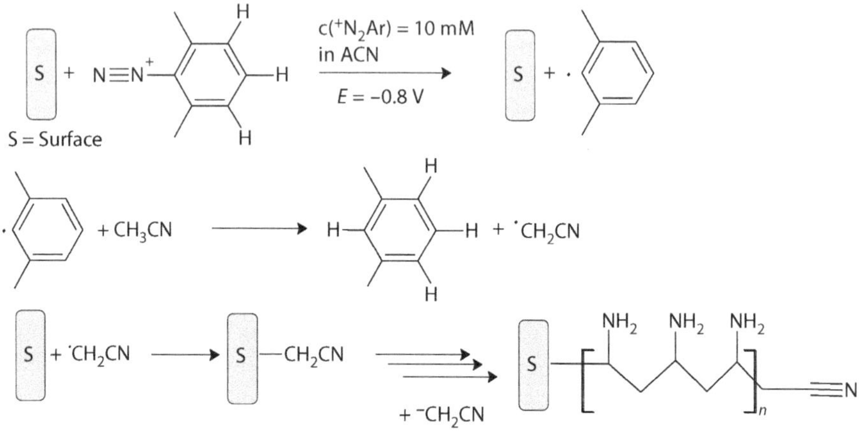

FIGURE 3.73   Indirect electrografting of ACN.

by reduction of the radical on the first grafted $-CH_2-CN$ groups, permits the growth of the chain up to 56 nm that is a polymer with 500 repeating units. If the $\cdot CH_2CN$ radical is obtained by thermal decomposition of benzoylperoxide, then the surface of hydrogenated diamond is modified only with $-CH_2C \equiv N$ groups. In this case, the cyanomethyl radical cannot be reduced to $^-CH_2CN$ [442].

A third example is provided by alkyl iodides [443] that are reduced at very negative potentials as discussed above. In the presence of **D(2,6-Me)2**, grafting takes place at $E_p = -0.25$ V/SCE with a positive potential shift of $\Delta V = 1.7$ V. In this case, modification of the surface also translates to a voltammogram that decreases in height upon repetitive scanning. However, in this case, the decreasing voltammogram is that of the diazonium (Figure 3.74), although the grafted species is derived from RI.

3.9.3   SPONTANEOUS, PHOTOCHEMICAL, OR CHEMICAL GRAFTING

Alkenes and alkynes can be grafted thermally or by photochemistry on Si–H [444–445] and diamond–H [446].

Azides generate the very reactive nitrenes upon irradiation. They react not only on carbon surfaces [447], but also on chalcogenide glasses [448] to give aziridine-modified surfaces. IN_3 in ACN reacts with PPF to give PPF-N3 surfaces [449].

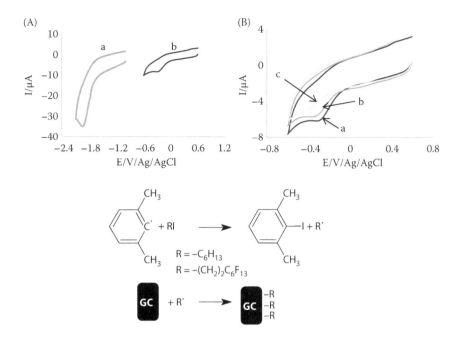

FIGURE 3.74 Cyclic voltammetry on a GC electrode in an ACN solution of (A) (a) I–$(CH_2)_2$–C_6F_{13}; (b) **D(2,6-Me)$_2$** recorded separately; and (B) recorded in the same solution where (a, b, c) refer to the 1st, 2nd, and 8th scan, respectively. (Adapted from Hetemi, D. et al. *Langmuir* 30, 2014: 13907–13913.)

Peroxides can be grafted photochemically on graphene. Electron transfer from photoexcited graphene to benzoyl peroxide was suggested to be the key step in this reaction [450].

With alkyne iodides, highly ordered SAMs having packing densities and molecular chain orientations very similar to those of alkanethiolates on Au(111) can be prepared by reaction of terminal *n*-alkynes: HC ≡ C(CH2)$_n$ CH$_3$, n = 5, 7, 9, and 11 with Au(111) at 60°C [451].

Thus, diazonium salts, thiol SAMs on gold, and silanes or phosphonic acids on oxidized surfaces have certainly been the most investigated methods. By adding the numerous different methods described in this section, it is possible to modify a very large number of surfaces to tailor at will their properties for specific applications.

3.10 CONCLUSION

From the preceding sections, it appears that electrografting or spontaneous grafting of diazonium salts and other species is now well understood, including the characteristics and the structures of the films as well as the grafting mechanism. For diazonium salts, some problems require a more definitive answer the spontaneous reaction on gold. However, further research should mainly be concerned with applications of electrografting in microelectronics, organic electronics, energy problems, analysis, sensors, biosensors, and theranostics. Industrial applications are now on the market such as stents, inks, paints, and more are under development [452].

REFERENCES

1. Delamar, M., R. Hitmi, J. Pinson, and J.-M. Savéant. Covalent modification of carbon surfaces by grafting of functionalized aryl radicals produced from electrochemical reduction of diazonium salts. *J. Am. Chem. Soc.* 114, 1992: 5883–5884.
2. Allongue, P., M. Delamar, B. Desbat, O. Fagebaume, R. Hitmi, J. Pinson, and J.-M. Savéant. Covalent modification of carbon surfaces by aryl radicals generated from the electrochemical reduction of diazonium salts. *J. Am. Chem. Soc.* 119, 1997: 201–207.
3. Pinson, J. and F. Podvorica. Attachment of organic layers to conductive or semiconductive surfaces by reduction of diazonium salts. *Chem. Soc. Rev.* 34, 2005: 429–439.
4. Bélanger, D. and J. Pinson. Electrografting: A powerful method for surface modification. *Chem. Soc. Rev.* 40, 2011: 3995–4048.
5. Gooding, J. J. and S. Ciampi. The molecular level modification of surfaces: From self-assembled monolayers to complex molecular assemblies. *Chem. Soc. Rev.* 40, 2011: 2704–2718.
6. Mahouche-Chergui, S., S. Gam-Derouich, C. Mangeney, and M. M. Chehimi. Aryl diazonium salts: A new class of coupling agents for bonding polymers, biomacromolecules and nanoparticles to surfaces. *Chem. Soc. Rev.* 40, 2011: 4143–4166.
7. Chehimi, M. M. Aryl diazonium Salts. *New Coupling Agents in Polymer and Surface Science.* Wiley-VCH, Weinheim, 2012.
8. McCreery, R. L. and A. J. Bergren. Surface functionalization in the nanoscale domain, in: *Nanofabrication: Techniques and Principles*, Maria Stepanova and Steven Dew (Eds.), Springer, New York, 2012, pp. 163–190.
9. Baranton, S. and D. Bélanger. Electrochemical derivatization of carbon surface by reduction of *in situ* generated diazonium cations. *J. Phys. Chem. B* 109, 2005: 24401–24410.
10. Lebègue, E., T. Brousse, J. Gaubicher, and C. Cougnon. Spontaneous arylation of activated carbon from aminobenzene organic acids as source of diazonium ions in mild conditions. *Electrochim. Acta* 88, 2013: 680–687.
11. Shul, G., R. Parent, H. A. Mosqueda, and D. Bélanger. Localized *in situ* generation of diazonium cations by electrocatalytic formation of a diazotization reagent. *ACS Appl. Mater. Interfaces* 5, 2013: 1468–1473.
12. Vinther, J., J. Iruthayaraj, K. Gothelf, S. U. Pedersen, and K. Daasbjerg. On electrogenerated acid-facilitated electrografting of aryltriazenes to create well-defined aryl-tethered films. *Langmuir* 29, 2013: 5181–5189.
13. Stewart, M. P., F. Maya, D. V. Kosynkin, S. M. Dirk, J. J. Stapleton, C. L. McGuiness, D. L. Allara, and J. M. Tour. Direct covalent grafting of conjugated molecules onto Si, GaAs, and Pd surfaces from aryldiazonium salts. *J. Am. Chem. Soc.* 126, 2004: 370–378.
14. Dyke, C. A. and J. M. Tour. Solvent-free functionalization of carbon nanotubes. *J. Am. Chem. Soc.* 125, 2003: 1156–1157.
15. Kölmel, D., K. N. Jung, and S. Bräse. Azides, diazonium ions, triazenes: Versatile nitrogen-rich functional groups. *Aust. J. Chem.* 67, 2014: 328–336.
16. Yu, Z.-q., Y.-w. Lv, C.-m. Yu, and W.-k. Su. Continuous flow reactor for Balz–Schiemann reaction: A new procedure for the preparation of aromatic fluorides. *Tetrahedron Lett.* 54, 2013: 1261–1263.
17. Galli, C. Evidence for the intermediacy of the aryl radical in the Sandmeyer reaction. *J. Chem. Soc. Perkin Trans.* 2, 1982: 1139–1141.

18. Kochi, J. K. The mechanism of the Sandmeyer and Meerwein reactions. *J. Am. Chem. Soc.* 79, 1957: 2942–2948.

19. Heinrich, M. R. Intermolecular olefin functionalisation involving aryl radicals generated from arenediazonium salts. *Chem. Eur. J.* 15, 2009: 820–833.

20. Chami, Z., M. Gareil, J. Pinson, J.-M. Savéant, and A. Thiébault. Aryl radicals from electrochemical reduction of aryl halides. Addition on olefins. *J. Org. Chem.* 56, 1991: 586–595.

21. Gomberg, M. and W. E. Bachmann. Synthesis of biaryl compounds by means of the diazo reaction. *J. Am. Chem. Soc.* 46, 1924: 2339–2343.

22. Bolton, R. and G. H. William. Homolytic arylation of aromatic and perfluoroaromatic compounds. *Chem. Soc. Rev.* 15, 1986: 61–289.

23. M'Halla, F., J. Pinson, and J. M. Savéant. The solvent as H-atom donor in organic electrochemical reactions. Reduction of aromatic halides. *J. Am. Chem. Soc.* 102, 1980: 4120–4127.

24. Berisha, A., C. Combellas, F. Kanoufi, J. Pinson, S. Ustaze, and F. I. Podvorica. Indirect grafting of acetonitrile-derived films on metallic substrates. *Chem. Mater.* 22, 2010: 2962–2969.

25. Sienkiewicz, A., M. Szymula, J. Narkiewicz-Michalek, and C. Bravo-Díaz. Formation of diazohydroxides ArN_2OH in aqueous acid solution: Polarographic determination of the equilibrium constant KR for the reaction of 4-substituted arenediazonium ions with H2O. *J. Phys. Org. Chem.* 27, 2014: 284–289.

26. Bravo Díaz, C. Diazohydroxides, diazoethers and related species, in: Patai's Chemistry of Functional Groups: The Chemistry of Hydroxylamines, Oximes and Hydroxamic Acids, Z. Rappoport, J. F. Liebman (Eds.), Vol. 2, p. 853, J. Wiley & Sons, Chichester, UK, 2011.

27. Pazo-Llorente, R., H. C. Bravo-Diaz, E. Gonzalez-Romero. Dediazoniation of 4-nitrobenzenediazonium ions in acidic $MeOH/H_2O$ mixtures: Role of acidity and MeOH concentration on the formation of transient diazo ethers that initiate homolytic dediazoniation. *Eur. J. Org. Chem.* 2006: 2201–2209.

28. Pazo-Llorente, R., C. Bravo-Díaz, E. González-Romero. Solvolysis of some arenediazonium salts in binary $EtOH/H_2O$ mixtures under acidic conditions. *Eur. J. Org. Chem.* 2003: 3421–3428.

29. Chen, B., A. K. Flatt, H. Jian, J. L. Hudson, and J. M. Tour. Molecular grafting to silicon surfaces in air using organic triazenes as stable diazonium sources and HF as a constant hydride-passivation source. *Chem. Mater.* 17, 2005: 4832–4836.

30. Elofson, R. M. The polarographic reduction of diazotized aromatic amines. *Can. J. Chem.* 36, 1958: 1207–1210.

31. Elofson, R. M. and F. F. Gadallah. Substituent effects in the polarography of aromatic diazonium salts. *J. Org. Chem.* 34, 1969: 854–857.

32. Atkinson, E. R., H. Warrenh, P. I. Abell, and R. E. Wing. A polarographic examination of diazotized amines. *J. Am. Chem. Soc.* 72, 1950: 915–918.

33. Bard, A. J., J. C. Gilbert, and R. D. Goodin. Application of spin trapping to the detection of radical intermediates in electrochemical transformations. *J. Am. Chem. Soc.* 96, 1974: 620–621.

34. Ahlberg, E., B. Helgée, and V. D. Parker. The reaction of aryl radicals with metallic electrodes. *Acta Chem. Scand. B* 34, 1980: 181–186.

35. Le Floch, F., G. Bidan, L. Pilan, E.-M. Ungureanu, and J.-P. Simonato. Carbon substrate functionalization with diazonium salts toward sensor applications. *Mol. Cryst. Liq. Cryst.* 486, 2008: 271[1313]–281[1323].

36. Shul, G., C. A. Castro Ruiz, D. Rochefort, P. A. Brooksby, D. Bélanger. Electrochemical functionalization of glassy carbon electrode by reduction of diazonium cations in protic ionic liquid. *Electrochim. Acta* 106, 2013: 378–385.

37. Benedetto, A., M. Balog, P. Viel, F. Le Derf, M. Sallé, S. Palacin. Electro-reduction of diazonium salts on gold: Why do we observe multi-peaks? *Electrochim. Acta* 53, 2008: 7117–7122.

38. Barba, F., B. Batanero, K. Tissaoui, N. Raouafi, K. Boujlel. Cathodic reduction of diazonium salts in aprotic medium. *Electrochem. Comm.* 12, 2010: 973–976.

39. Gan, L., D. Zhang, and X. Guo. Electrochemistry: An efficient way to chemically modify individual monolayers of graphene. *Small* 8, 2012: 1326–1330.

40. Clausmeyer, J., J. Henig, W. Schuhmann, and N. Plumeré. Scanning droplet cell for chemoselective patterning through local electroactivation of protected quinone monolayers. *Chem. Phys. Chem.* 15, 2014: 151–156.

41. Berisha, A., C. Combellas, F. Kanoufi, J. Pinson, F. I. Podvorica. Physisorption vs grafting of aryldiazonium salts onto iron: A corrosion study. *Electrochim. Acta* 56, 2011: 10762–10766.

42. Ceccato, M., A. Bousquet, M. Hinge, S. U. Pedersen, and K. Daasbjerg. Using a mediating effect in the electroreduction of aryldiazonium salts to prepare conducting organic films of high thickness. *Chem. Mater.* 23, 2011: 1551–1557.

43. Cougnon, C., F. Gohier, D. Bélanger, J. Mauzeroll. *In Situ* formation of diazonium salts from nitro precursors for scanning electrochemical microscopy patterning of surfaces. *Ang. Chem. Int. Ed.* 48, 2009: 4006–4009.

44. Jayasundara, D. R., R. J. Cullen, L. Soldi, and P. E. Colavita. *In situ* studies of the adsorption kinetics of 4-nitrobenzenediazonium salt on gold. *Langmuir* 27, 2011: 13029–13036.

45. Andrieux, C. P. and J. Pinson. The standard redox potential of the phenyl radical/anion couple. *J. Am. Chem. Soc.* 125, 2003: 14801–14806.

46. Vase, K. Novel approaches for electrochemically assisted covalent modification of carbon surfaces, PhD Thesis, University of Aarhus, DK. 2007.

47. Lin, C. Y., M. L. Coote, A. Gennaro, and K. Matyjaszewski. Ab initio evaluation of the thermodynamic and electrochemical properties of alkyl halides and radicals and their mechanistic implications for atom transfer radical polymerization. *J. Am. Chem. Soc.* 130, 2008: 12762–1277.

48. Fu, Y., L. Liu, H.-Z. Yu, Y.-M. Wang, and Q.-X. Guo. Quantum-chemical predictions of absolute standard redox potentials of diverse organic molecules and free radicals in acetonitrile. *J. Am. Chem. Soc.* 127, 2005: 7227–7234.

49. Romańczyk, P. P., G. Rotko, S. S. Kurek. The redox potential of the phenyl radical/ anion couple and the effect thereon of the lithium cation: A computational study. *Electrochem. Comm.* 48, 2014: 21–23.

50. Fau, M., A. Kowalczyk, P. Olejnik, and A. M. Nowicka. Tight and uniform layer of covalently bound aminoethylophenyl groups perpendicular to gold surface for attachment of biomolecules. *Anal. Chem.* 83, 2011: 9281–9288.

51. Barrière, F., and A. J. Downard. Covalent modification of graphitic carbon substrates by non-electrochemical methods. *J Solid State Electrochem.* 12, 2008: 1231–1244.

52. Mesnage, A., G. Deniau, and S. Palacin. Grafting polyphenyl-like films on metallic surfaces using galvanic anodes. *RSC Advances* 3, 2013: 13901–13906.

53. Adenier, A., E. Cabet-Deliry, A. Chaussé, S. Griveau, F. Mercier, J. Pinson, and C. Vautrin-Ul. Grafting of nitrophenyl groups on carbon and metallic surfaces without electrochemical induction. *Chem. Mater.* 17, 2005: 491–501.

54. Adenier, A., N. Barré, E. Cabet-Deliry, A. Chaussé, S. Griveau, F. Mercier, J. Pinson, and C. Vautrin-Ul. Study of the spontaneous organic layers on carbon and metal surfaces from diazonium salts. *Surf. Sci.* 600, 2006: 4801–4812.

55. Chamoulaud, G. and D. Bélanger. Spontaneous derivatization of a copper electrode with *in situ* generated diazonium cations in aprotic and aqueous media. *J. Phys. Chem. C* 111, 2007: 7501–7507.

56. Hurley, B. L. and R. L. McCreery. Covalent bonding of organic molecules to Cu and Al alloy 2024 T3 surfaces via diazonium ion reduction. *J. Electrochem. Soc.* 151, 2004: B252–B259.

57. Hapiot, P. and J. Simonet. Reduction of platinum under superdry conditions, an electrochemical approach. *Electroanal. Chem.* 23, 2010: 105–169.

58. Ghilane, J. and J.-C. Lacroix. Formation of a bifunctional redox system using electrochemical reduction of platinum in ferrocene based ionic liquid and its reactivity with aryldiazonium. *J. Am. Chem. Soc.* 135, 2013: 4722–4728.

59. Toupin, M. and D. Bélanger. Spontaneous functionalization of carbon black by reaction with 4-nitrophenyldiazonium cations. *Langmuir* 24, 2008: 1910–1917.

60. LeFloch, F., J.-P. Simonato, and G. Bidan. Electrochemical signature of the grafting of diazonium salts: A probing parameter for monitoring the electro addressed functionalization of devices. *Electrochim. Acta* 54, 2009: 3078–3085.

61. Lehr, J., B. E. Williamson, and A. J. Downard. Spontaneous grafting of nitrophenyl groups to planar glassy carbon substrates: Evidence for two mechanisms. *J. Phys. Chem. C* 115, 2011: 6629–6634.

62. Cullen, R. J., D. R. Jayasundara, L. Soldi, J. J. Cheng, G. Dufaure, and P. E. Colavita. Spontaneous grafting of nitrophenyl groups on amorphous carbon thin films: A structure – reactivity investigation. *Chem. Mater.* 24, 2012: 1031–1040.

63. Chehimi, M. M., J. Pinson, and Z. Salmi. Carbon nanotubes. In *Surface modification and applications in Applied Surface Chemistry of Nanomaterials,* M. M. Chehimi, J. Pinson (Eds.), Nova Science Publishers Inc., New York, 2013, pp. 95–144.

64. Dyke, C. A., M. P. Stewart, F. Maya, and J. M. Tour. Diazonium-based functionalization of carbon nanotubes: XPS and GC-MS analysis and mechanistic implications. *Synlett,* 2004: 155–160.

65. Schmidt, G., S. Gallon, S. Esnouf, J.-P. Bourgoin, and P. Chenevier. Mechanism of the coupling of diazonium to single-walled carbon nanotubes and its consequences. *Chem. Eur. J.* 15, 2009: 2101–2110.

66. Ménard-Moyon, C. Chemistry of graphene, in: *Applied Chemistry of Nanomaterials.* M. M. Chehimi, J. Pinson (Eds.), Nova Science Publishers Inc., New York, 2013, pp. 57–94.

67. Bekyarova, E., M. E. Itkis, P. Ramesh, C. Berger, M. Sprinkle, W. A. deHeer, and R. C. Haddon. Chemical modification of epitaxial graphene: Spontaneous grafting of aryl groups. *J. Am. Chem. Soc.* 131, 2009: 1336–1337.

68. Koehler, F. M., A. Jacobsen, K. Ensslin, C. Stampfer, and W. J. Stark. Selective chemical modification of graphene surfaces: Distinction between single- and bilayer graphene. *Small* 6, 2010: 1125–1130.

69. Wang, Q. H., Z. Jin, K. K. Kim, A. J. Hilmer, G. L. C. Paulus, C.-J. Shih, M.-H. Ham et al. Understanding and controlling the substrate effect on graphene electron-transfer chemistry via reactivity imprint lithography. *Nature Chem.* 4, 2012: 724–732.

70. Hirsch, A., J. M. Englert, and F. Hauke. Wet chemical functionalization of graphene. *Acc. Chem. Res.* 46, 2013: 87–96.

71. Charlier, J., E. Clolus, C. Bureau, and S. Palacin. Selectivity of organic grafting as a function of the nature of semiconducting substrates. *J. Electroanal. Chem.* 625, 2009: 97–100.

72. Chulki, K., O. Kiwon, H. Seunghee, K. Kyungkon, K. Il Won, and K. Heesuk. Electroless chemical grafting of nitrophenyl groups on *n*-doped hydrogenated amorphous silicon surfaces. *J. Nanosci. Nanotechnol.* 14, 2014: 6309–6313.

73. Girard, A., F. Geneste, N. Coulon, C. Cardinaud, and T. Mohammed-Brahim. SiGe derivatization by spontaneous reduction of aryl diazonium salts. *Appl. Surf. Sci.* 282, 2013: 146–155.

74. Lefevre, X., O. Segut, P. Jegou, S. Palacin, and B. Jousselme. Towards organic film passivation of germanium wafers using diazonium salts: Mechanism and ambient stability. *Chem. Sci.* 3, 2012: 1662–1667.

75. Collins, G., P. Fleming, C. O'Dwyer, M. A. Morris, and J. D. Homes. Organic functionalization of germanium nanowires using arenediazonium salts. *Chem Mater.* 23, 2011: 1883–1891.

76. Mesnage, A, S. Esnouf, P. Jégou, G. Deniau, and S. Palacin. Understanding the redox-induced polymer grafting process: A dual surface-solution analysis. *Chem. Mater.* 22, 2010: 6229–6239.

77. Poobalasingam, A., G. G. Wildgoose, and R. G. Compton. A mechanistic investigation into the covalent chemical derivatisation of graphite and glassy carbon surfaces using aryldiazonium salts. *J. Phys. Org. Chem.* 21, 2008: 433–439.

78. Simons, B. M., J. Lehr, D. J. Garrett, A. J. Downward. Formation of thick aminophenyl films from aminobenzenediazonium ion in the absence of a reduction source. *Langmuir* 30, 2014: 4989–4996.

79. Mévellec, V., S. Roussel, L. Tessier, J. Chancolon, M. Mayne-L'Hermite, G. Deniau, P. Viel, and S. Palacin. Grafting polymers on surfaces: A new powerful and versatile diazonium salt-based one-step process in aqueous media. *Chem. Mater.* 19, 2007: 6323–6330.

80. Busson, M., A. Berisha, C. Combellas, F. Kanoufi, and J. Pinson. Photochemical grafting of diazonium salts on metals. *Chem. Commun.* 47, 2011: 12631–12633.

81. Kosynkin, D., T. M. Bockman, and J. K. Kochi. Thermal (iodide) and photoinduced electron-transfer catalysis in biaryl synthesis via aromatic arylations with diazonium salts. *J. Am. Chem. Soc.* 119, 1997: 4846–4855.

82. Bouriga, M., M. M. Chehimi, C. Combellas, P. Decorse, F. Kanoufi, A. Deronzier, and J. Pinson. Sensitized photografting of diazonium salts by visible light. *Chem. Mater.* 25, 2013: 90–97.

83. Cano-Yelo, H. and A. Deronzier. Photocatalysis of the Pschorr reaction by tris-(2,2'-bipyridyl) ruthenium (11) in the phenanthrene series. *J. Chem. Soc. Perkin Trans. II*, 1984: 1093–1098.

84. Garcia, A., N. Hanifi, B. Jousselme, P. Jégou, S. Palacin, P. Viel, and T. Berthelot. Polymer grafting by inkjet printing: A direct chemical writing toolset. *Adv. Funct. Mater.* 23, 2013: 3668–3674.

85. Verberne-Sutton, S. D., R. D. Quarels, X. Zhai, J. C. Garno, and J. R. Ragains. Application of visible light photocatalysis with particle lithography to generate polynitrophenylene nanostructures. *J. Am. Chem. Soc.* 136, 2014: 14438–14444.

86. Mirkhalaf, F., T. J. Mason, D. J. Morgan, and V. Saez frequency. Effects on the surface coverage of nitrophenyl films ultrasonically grafted onto indium tin oxide. *Langmuir* 27, 2011: 1853–1858.

87. Fernàndez-Alonso, A. and C. Bravo-Dìaz. Methanolysis of 4-bromobenzenediazonium ions. Effects of acidity, [MeOH] and temperature on the formation and decomposition of diazoethers that initiate homolytic dediazoniation. *Org. Biomol. Chem* 6, 2008: 4004–4011.

88. Podvorica, F. I., F. Kanoufi, J. Pinson, and C. Combellas. Spontaneous grafting of diazoates on metals. *Electrochim. Acta* 54, 2009: 2164–2170.

89. Griffete, N., R. Ahmad, H. Benmehdi, A. Lamouri, P. Decorse, and C. Mangeney. Elaboration of hybrid silica particles using a diazonium salt chemistry approach. *Coll. Surf. A* 439, 2013: 145–150.

90. Chehimi, M. M., A. Lamouri, M. Picot, and J. Pinson. Surface modification of polymers by reduction of diazonium salts: Polymethylmethacrylate as an example. *J. Mater. Chem. C* 2, 2014: 356–363.

91. Liu, J., M. R. Zubiri, B. Vigolo, M. Dossot, B. Humbert, Y. Fort, and E. McRae. Microwave-assisted functionalization of single-wall carbon nanotubes through diazonium. *J. Nanosci. Nanotechnol.* 7, 2007: 3519–3523.

92. Rubio, N., A. Herrero, M. Meneghetti, A. Diaz—Ortiz, M. Schiavon, M. Prato, and E. Vazquez. Efficient functionalization of carbon nanohorns via microwave irradiation. *J. Mater. Chem.* 19, 2009: 4407–4413.

93. Brunetti, F. G., M. A. Herrero, J. de M. Munoz, A. Diaz-Ortiz, J. Alfonsi, M. Meneghetti, M. Prato, and E. Vazquez. Microwave-induced multiple functionalization of carbon nanotubes. *J. Am. Chem. Soc.* 130, 2008: 8094–8100.

94. Cui, B., J.-Y. Gu, T. Chen, H.-J. Yan, D. Wang, and L.-J. Wan. Solution effect on diazonium-modified Au(111): Reactions and structures. *Langmuir* 29, 2013: 2955–2960.

95. Jousselme, B., G. Bidan, M. Billon, C. Goyer, Y. Kervella, S. Guillerez, E. Abou Hamad, C. Goze-Bac, and J.-Y. Mevellec. One-step electrochemical modification of carbon nanotubes by ruthenium complexes via new diazonium salts. *J. Electroanal. Chem.* 621, 2008: 277–285.

96. Descroix, S., G. Hallais, C. Lagrost, and J. Pinson. Regular poly(para-phenylene) films bound to gold surfaces through the electrochemical reduction of diazonium salts followed by electropolymerization in an ionic liquid. *Electrochim. Acta* 106, 2013: 172–180.

97. Schlenoff, J. B., M. Li, and H. Ly. Stability and self-exchange in alkanethiol monolayers. *J. Am. Chem. Soc.* 117, 1995: 12528–12536.

98. Shewchuk, D. M. and M. T. McDermott. Comparison of diazonium salt derived and thiol derived nitrobenzene layers on gold. *Langmuir* 25, 2009: 4556–4563.

99. Ghorbal, A., F. Grisotto, M. Laudé, J. Charlier, and S. Palacin. The *in situ* characterization and structuring of electrografted polyphenylene films on silicon surfaces. An AFM and XPS study. *J. Coll. Interf. Sci.* 328, 2008: 308–313.

100. Henry de Villeneuve, C., J. Pinson, M. C. Bernard, and P. Allongue. Electrochemical formation of close-packed phenyl layers on Si(111). *J. Phys. Chem. B* 101, 1997: 2415–2420.

101. D'Amours, M. and D. Bélanger. Stability of substituted phenyl groups electrochemically grafted at carbon electrode Surface. *J. Phys. Chem. B* 107, 2003: 4811–4817.

102. Cortès, E., A. A. Rubert, G. Benitez, P. Carro, M. E. Vela, and R. C. Salvarezza. Enhanced stability of thiolate self-assembled monolayers (SAMs) on nanostructured gold substrates. *Langmuir* 25, 2009: 5661–5666.

103. Holm, A. H., R. Møller, K. H. Vase, M. Dong, K. Norrman, F. Besenbacher, S U. Pedersen, and K. Daasbjerg. Nucleophilic and electrophilic displacements on covalently modified carbon: Introducing 4,4'-bipyridinium on grafted glassy carbon electrodes. *New J. Chem.* 29, 2005: 659–666.

104. Toupin, M. and D. Bélanger. Thermal stability study of aryl modified carbon black by *in situ* generated diazonium salt. *J. Phys. Chem. C* 111, 2007: 5394–5540.

105. Dyke, C. A. and J. M. Tour. Unbundled and highly functionalized carbon nanotubes from aqueous reactions. *Nano Lett.* 3, 2003: 1215–1218.

106. Dyke, C. A. and J. M. Tour. Overcoming the insolubility of carbon nanotubes through high degrees of sidewall functionalization. *Chem. Eur. J.* 10, 2004: 812–817.

107. Hinge, M., E. S. Gonçalves, S. U. Pedersen, and K. Daasbjerg. On the electrografting of stainless steel from para-substituted aryldiazonium salts and the thermal stability of the grafted layer. *Surf. Coatings Technol.* 205, 2010: 820–827.

108. Aït Atmane, Y., L. Sicard, A. Lamouri, J. Pinson, M. Sicard, C. Masson, S. Nowak et al. Functionalization of aluminum nanoparticles using a combination of aryl diazonium salt chemistry and iniferter method. *J. Phys. Chem. C* 117, 2013: 26000–26006.

109. Valiokas, R., M. O. R. Østblom, S. Svedhem, S. C. T. Svensson, and B. Liedberg. Thermal stability of self-assembled monolayers: Influence of lateral hydrogen bonding. *J. Phys. Chem. B* 106, 2002: 10401–10409.

110. Boukerma, K., M. M. Chehimi, J. Pinson, and C. Blomfield. X-ray photoelectron spectroscopy evidence for the covalent bond between an iron surface and aryl groups attached by the electrochemical reduction of diazonium salts. *Langmuir* 19, 2003: 6333–6335.

111. Bell, K. J., P. A. Brooksby, M. I. J. Polson, and A. J. Downard. Evidence for covalent bonding of aryl groups to MnO2 nanorods from diazonium-based grafting. *Chem. Comm.* 50, 2014: 13687–13690.

112. Mesnage, A., X. Lefèvre, P. Jégou, G. Deniau, and S. Palacin. Spontaneous grafting of diazonium salts: Chemical mechanism on metallic surfaces. *Langmuir* 28, 2012: 11767–11778.

113. Laforgue, A., T. Addou, and D. Bélanger. Characterization of the deposition of organic molecules at the surface of gold by the electrochemical reduction of aryldiazonium cations. *Langmuir* 21, 2005: 6855–6865.

114. Le, X. T., G. Zeb, P. Jégou, and T. Berthelot. Electrografting of stainless steel by the diazonium salt of 4-aminobenzylphosphonic acid. *Electrochim. Acta* 71, 2012: 66–72.

115. Kawai, K., T. Narushima, K. Kaneko, H. Kawakami, M. Matsumoto, A. Hyono, H. Nishihara, and T. Yonezawa. Synthesis and antibacterial properties of water-dispersible silver nanoparticles stabilized by metal–carbon σ-bonds. *Appl. Surf. Sci.* 262, 2012: 76–80.

116. Lyskawa, J. and D. Bélanger. Direct modification of a gold electrode with aminophenyl groups by electrochemical reduction of *in situ* generated aminophenylmonodiazonium cations. *Chem. Mater.* 18, 2006: 4755–4763.

117. Kullapere, M., J. Kozlova, L. Matisen, V. Sammelselg, H. A. Menezes, G. Maia, D. J. Schiffrin, and K. Tammeveski. Electrochemical properties of aryl-modified gold electrodes. *J. Electroanal. Chem.* 641, 2010: 90–98.

118. Salmi, Z., S. Gam-Derouich, S. Mahouche-Chergui, M. Turmine, M. M. Chehimi. On the interfacial chemistry of aryl diazonium compounds in polymer science. *Chem. Papers* 66, 2012: 369–391.

119. Miyachi, M., Y. Yamamoto, Y. Yamanoi, A. Minoda, S. Oshima, Y. Kobori, and H. Nishihara. Synthesis of diazenido-ligated vanadium nanoparticles. *Chem. Mater.* 18, 2006: 4755–4763.

120. Ricci, A. M., L. P. Méndez De Leo, F. J. Williams, and E. J. Calvo. Some evidence for the formation of an azo bond during the electroreduction of diazonium salts on Au substrates. *Chem. Phys. Chem.* 13, 2012: 2119–2127.

121. Samanta, S., I. Bakas, A. Singh, D. K. Aswal, and M. M. Chehimi. *In situ* diazonium-modified flexible ITO-coated PEN substrates for the deposition of adherent silver-polypyrrole nanocomposite films. *Langmuir* 30, 2014: 9397–9406.

122. Quinton, D., A. Galtayries, F. Prima, and S. Griveau. Functionalization of titanium surfaces with a simple electrochemical strategy. *Surf. Coating. Technol.* 206, 2012: 2302–2307.

123. Combellas, C., F. Kanoufi, J. Pinson, and F. I. Podvorica. Time-of-flight secondary ion mass spectroscopy characterization of the covalent bonding between a carbon surface and aryl groups. *Langmuir* 21, 2005: 280–286.

124. Piper, D. J. E., G. J. Barbante, N. Brack, P. J. Pigram, and C. F. Hogan. Highly stable ECL active films formed by the electrografting of a diazotized ruthenium complex generated *in situ* from the amine. *Langmuir* 27, 2011: 474–480.

125. Ahmad, R., L. Boubekeur-Lecaque, M. Nguyen, S. Lau-Truong, A. Lamouri, P. Decorse, A. Galtayries, J. Pinson, N. Felidj, and C. Mangeney. Tailoring the surface chemistry of gold nanorods through Au – C/Ag – C covalent bonds using aryl diazonium salts. *J. Phys. Chem. C* 118, 2014: 19098–19105.

126. Doppelt, P., G. Hallais, J. Pinson, F. Podvorica, and S. Verneyre. Surface modification of conducting substrates. Existence of azo bonds in the structure of organic layers obtained from diazonium salts. *Chem. Mater.* 19, 2007: 4570–4575.

127. Itoh, T. and R. L. McCreery. *In situ* Raman spectroelectrochemistry of electron transfer between glassy carbon and a chemisorbed nitroazobenzene monolayer. *J. Am. Chem. Soc.* 124, 2002: 10894–10902.

128. Laurentius, L., S. R. Stoyanov, S. Gusarov, A. Kovalenko, R. Du, G. P. Lopinski, and M. T. McDermott. Diazonium-derived aryl films on gold nanoparticles: Evidence for a carbon gold covalent bond. *ACS Nano* 5, 2011: 4219–4227.

129. Huang, P., L. Jing, H. Zhu, and X. Gao. Diazonium functionalized graphene: Microstructure, electric, and magnetic properties. *Acc. Chem. Res.* 46, 2013: 43–52.

130. Jiang, D.-e., B. G. Sumpter, and S. Dai. Structure and bonding between an aryl group and metal surfaces. *J. Am. Chem. Soc.* 128, 2006: 6030–6031.

131. de la Llave, E., A. Ricci, E. J. Calvo, and D. A. Scherlis. Binding between carbon and the Au(111) surface and what makes it different from the S-Au(111) bond. *J. Phys. Chem. C* 112, 2008: 17611–17617.

132. Zhao, J.-x. and Y.-h. Ding. Chemical functionalization of single-walled carbon nanotubes (SWNTs) by aryl groups: A density functional theory study. *J. Phys. Chem. C* 112, 2008: 13141–13149.

133. Sumpter, B. G., D.-e. Jiang, and V. Meunier. New insight into carbon-nanotube electronic-structure selectivity. *Small* 4, 2008: 2035–2042.

134. Laflamme Janssen, J., J. Beaudin, N. D. M. Hine, P. D. Haynes, and M. Côté. Bromophenyl functionalization of carbon nanotubes: An ab initio study. *Nanotechnology* 24, 2013: 375702 (8pp).

135. Jiang, D.-e., B. G. Sumpter, and S. Dai. How do aryl groups attach to a graphene sheet? *J. Phys. Chem. B* 110, 2006: 22628–22632.

136. Kariuki, J. K. and M. T. McDermott. Nucleation and growth of functionalized aryl films on graphite electrodes. *Langmuir* 15, 1999: 6534–6540.

137. Ma, H., L. Lee, P. A. Brooksby, S. A. Brown, S. J. Fraser, K. C. Gordon, Y. R. Leroux, P. Hapiot, and A. J. Downard. Scanning tunneling and atomic force microscopy evidence for covalent and noncovalent interactions between aryl films and highly ordered pyrolytic graphite. *J. Phys. Chem. C* 118, 2014: 5820 – 5826.

138. Kirkman, P. M., A. G. Guell, A. S. Cuharuc, and P. R. Unwin. Spatial and temporal control of the diazonium modification of sp2 carbon surfaces. *J. Am. Chem. Soc.* 136, 2014: 36–39.

139. Lim, H., J. S. Lee, H.-J. Shin, H. S. Shin, and H. C. Choi. Spatially resolved spontaneous reactivity of diazonium salt on edge and basal plane of graphene without surfactant and its doping effect. *Langmuir* 26, 2010: 12278–12284.

140. Wang, Q. H., C.-J. Shih, G. L. C. Paulus, and M. S. Strano. Evolution of physical and electronic structures of bilayer graphene upon chemical functionalization. *J. Am. Chem. Soc.* 135, 2013: 18866–18875.

141. Combellas, C., F. Kanoufi, J. Pinson, and F. I. Podvorica. Sterically hindered diazonium salts for the grafting of a monolayer on metals. *J. Am. Chem. Soc.* 130, 2008: 8576–8577.

142. Greenwood, J., T. H. Phan, Y. Fujita, Z. Li, O. Ivasenko, W. Vanderlinden, H. Van Gorp, W. Frederickx et al. Covalent modification of graphene and graphite using diazonium chemistry: Tunable grafting and nanomanipulation. *ACS Nano* 9, 2015: 5520–5535.

143. Fontaine, O., J. Ghilane, P. Martin, J.-C. Lacroix, and H. Randriamahazaka. Ionic liquid viscosity effects on the functionalization of electrode material through the electroreduction of diazonium salts. *Langmuir* 26, 2010: 18542–18549.

144. Chen, X., M. Chockalingam, G. Liu, E. Luais, A. L. Gui, and J. J. Gooding. A molecule with dual functionality 4-aminophenylmethylphosphonic acid: A comparison between layers formed on indium tin oxide by *in situ* generation of an aryl diazonium salt or by self-assembly of the phosphonic acid. *Electroanalysis* 23, 2011: 2633–2642.

145. Liu, G., J. Liu, T. Böcking, P. K. Eggers, and J. J. Gooding. The modification of glassy carbon and gold electrodes with aryl diazonium salt: The impact of the electrode materials on the rate of heterogeneous electron transfer. *Chem. Phys.* 319, 2005: 136–146.

146. Griveau, S., D. Mercier, C. Vautrin-Ul, and A. Chaussé. Electrochemical grafting by reduction of 4-aminoethylbenzenediazonium salt: Application to the immobilization of (bio)molecules. *Electrochem. Comm.* 9, 2007: 2768–2773.

147. Hauquier, F., N. Debou, S. Palacin, and B. Jousselme. Amino functionalized thin films prepared from Gabriel synthesis applied on electrografted diazonium salts. *J. Electroanal. Chem.* 677–680, 2012: 127–132.

148. Lee, L., P. A. Brooksby, and A. J. Downard. The stability of diazonium ion terminated films on glassy carbon and gold electrodes. *Electrochem. Comm.* 19, 2012: 67–69.

149. Martin, C., M. Alias, F. Christien, O. Crosnier, D. Bélanger, and T. Brousse. Graphite-grafted silicon nanocomposite as a negative electrode for lithium-ion batteries. *Advanced Materials* 21, 2009: 4735–4741.

150. Viel, P., X. T. Le, V. Huc, J. Bar, A. Benedetto, A. Le Goff, A. Filoramo et al. Covalent grafting onto self-adhesive surfaces based on aryldiazonium salt seed layers. *J. Mater. Chem.* 18, 2008: 5913–5920.

151. Liu, G Z., E. Luais, and J. J. Gooding. The fabrication of stable gold nanoparticle-modified interfaces for electrochemistry. *Langmuir* 27, 2011: 4176–4183.

152. Radi, A. E., X. Munoz-Berbel, M. Cortina-Ping, and J. L. Marty. Novel protocol for covalent immobilization of horseradish peroxidase on gold electrode surface. *Electroanalysis* 21, 2009: 696–700.

153. Li, X, X. Wang, G. Ye, W. Xia, and X. Wang. Polystyrene-based diazonium salt as adhesive: A new approach for enzyme immobilization on polymeric supports. *Polymer* 51, 2010: 860–867.

154. Flatt, A. K., B. Chen, and J. M. Tour. Fabrication of carbon nanotube – molecule – silicon junctions. *J. Am. Chem. Soc.* 127, 2005: 8918–8919.

155. de Fuentes, O. A., T. Ferri, M. Frasconi, V. Paolini, and R. Santucci. Highly-ordered covalent anchoring of carbon nanotubes on electrode surfaces by diazonium salt reactions. *Angew. Chem. Int. Ed.* 50, 2011: 3457–3461.

156. Joyeux, X., P. Mangiagalli, and J. Pinson. Localized attachment of carbon nanotubes in microelectronic structures. *Adv. Mat.* 21, 2009: 4404–4408.

157. Martin, C., O. Crosnier, R. Retoux, D. Bélanger, D. M. Schleich, and T. Brousse. Chemical coupling of carbon nanotubes and silicon nanoparticles for improved negative electrode performance in lithium-ion batteries. *Adv. Func. Mater.* 21, 2011: 3524–3530.

158. Marshall, N. and J. Locklin. Reductive electrografting of benzene (*p*-bisdiazonium hexafluorophosphate): A simple and effective protocol for creating diazonium-functionalized thin films. *Langmuir* 27, 2011: 13367–13373.

159. Agullo, J., S. Canesi, F. Schaper, M. Morin, and D. Bélanger. Formation and reactivity of 3-diazopyridinium cations and influence on their reductive electrografting on glassy carbon. *Langmuir* 28, 2012: 4889–4895.

160. Viel, P., J. Walter, S. Bellon, and T. Berthelot. Versatile and nondestructive photochemical process for biomolecule immobilization. *Langmuir* 29, 2013: 2075–2082.

161. Lin, S., C.-W. Lin, J.-H. Jhang, and W.-H. Hung. Electrodeposition of long-chain alkyl-aryl layers on Au surfaces. *J. Phys. Chem. C* 116, 2012: 17048–17054.

162. Fairman, C., J. Z. Ginges, S. B. Lowe, and J. J. Gooding. Protein resistance of surfaces modified with oligo(ethylene glycol) aryl diazonium derivatives. *Chem. Phys. Chem.* 14, 2013: 2183–2189.

163. Jayasundara, D. R., T. Duff, M. D. Angione, J. Bourke, D. M. Murphy, E. M. Scanlan, and P. E. Colavita. Carbohydrate coatings via aryldiazonium chemistry for surface biomimicry. *Chem. Mater.* 25, 2013: 4122–4128.

164. Cougnon, C., S. Boisard, O. Cador, M. Dias, E. Levillain, and T. Breton. A facile route to steady redox-modulated nitroxide spin-labeled surfaces based on diazonium chemistry. *Chem. Commun.* 49, 2013: 4555–4557.

165. Blinco, J. P., B. A. Chalmers, A. Chou, K. E. Fairfull-Smith, and S. E. Bottle. Spin-coated carbon. *Chem. Sci.* 4, 2013: 3411–3415.
166. Cannizzo, C., M. Wagner, J.-P. Jasmin, C. Vautrin-Ul, D. Doizi, C. Lamouroux, and A. Chaussé. Calix[6]arene mono-diazonium salt synthesis and covalent immobilization onto glassy carbon electrodes. *Tetrahedron Lett.* 55, 2014: 4315–4318.
167. Mattiuzzi, A., I. Jabin, C. Mangeney, C. Roux, O. Reinaud, L. Santos, J.-F. Bergamini, P. Hapiot, and C. Lagrost. Electrografting of calix[4]arenediazonium salts to form versatile robust platforms for spatially controlled surface functionalization. *Nature Comm.* 3, 2012: 1130–1138.
168. Picot, M., I. Nicolas, C. Poriel, J. Rault-Berthelot, and F. Barrière. On the nature of the electrode surface modification by cathodic reduction of tetraarylporphyrin diazonium salts in aqueous media. *Electrochem. Comm.* 20, 2012: 167–170.
169. Gómez-Anquela, C., M. Revenga-Parra, J. M. Abad, A. García Marín, J. L. Pau, F. Pariente, J. Piqueras, and E. Lorenzo. Electrografting of N',N'-dimethylphenothiazin-5-ium-3,7-diamine (Azure A) diazonium salt forming electrocatalytic organic films on gold or graphene oxide gold hybrid electrodes. *Electrochim. Acta* 116, 2014: 59–68.
170. Kibena, E., M. Marandi, U. Mäeorg, L. B. Venarusso, G. Maia, L. Matisen, A. Kasikov, V. Sammelselg, and K. Tammeveski. Electrochemical modification of gold electrodes with azobenzene derivatives by diazonium reduction. *Chem. Phys. Chem.* 14, 2013: 1043–1054.
171. Guo, K., X. Chen, S. Freguia, and B. C. Donose. Spontaneous modification of carbon surface with neutral red from its diazonium salts for bioelectrochemical systems. *Biosens. Bioelec.* 47, 2013: 184–189.
172. Rinfray, C., G. Izzet, J. Pinson, S. Gam Derouich, J.-J. Ganem, C. Combellas, F. Kanoufi, and A. Proust. Electrografting of diazonium-functionalized polyoxometalates: Synthesis, immobilisation and electron-transfer characterisation from glassy carbon. *Chem. Eur. J.* 19, 2013: 13838–13846.
173. Gam Derouich, S., C. Rinfray, G. Izzet, J. Pinson, J.-J. Gallet, F. Kanoufi, A. Proust, and C. Combellas. Control of the grafting of hybrid polyoxometalates on metal and carbon surfaces: Toward submonolayers. *Langmuir* 30, 2014: 2287–2296.
174. Jacques, A., S. Devillers, J. Delhalle, and Z. Mekhalif. Electrografting of in situ generated pyrrole derivative diazonium salt for the surface modification of nickel. *Electrochim. Acta* 109, 2013: 781–789.
175. Martin, P., M. L. Della Rocca, A. Anthore, P. Lafarge, and J.-C. Lacroix. Organic electrodes based on grafted oligothiophene units in ultrathin, large-area molecular junctions. *J. Am. Chem. Soc.* 134, 2012: 154–157.
176. Polsky, R., J. C. Harper, D. R. Wheeler, S. M. Dirk, D. C. Arango, and S. M. Brozik. Electrically addressable diazonium-functionalized antibodies for multianalyte electrochemical sensor applications. *Biosens. Bioelec.* 23, 2008: 757–764.
177. Marquette, C. A., B. P. Corgier, K. A. Heyries, and L. J. Blum. Biochips: Non-conventional strategies for biosensing elements immobilization. *Front. Biosci.* 13, 2008: 382–400.
178. Combellas, C., D.-e. Jiang, F. Kanoufi, J. Pinson, and F. I. Podvorica. Steric effects in the reaction of aryl radicals on surfaces. *Langmuir* 25, 2009: 286–293.
179. Zhang, X., F. Rösicke, V. Syritski, G. Sun, J. Reut, K. Hinrichs, S. Janietz, and J. Rappich. Influence of the para-substitutent of benzene diazonium salts and the solvent on the film growth during electrochemical reduction. *Z. Phys. Chem.* 228, 2014: 557–573.
180. M. M. Chehimi, J. Pinson (Eds.), *Applied Surface Chemistry of Nanomaterials.* Nova Science Publishers Inc., New York, 2013.
181. Kuo, T.-C., R. L. McCreery, and G. M. Swain. Electrochemical modification of boron-doped chemical vapor deposited diamond surfaces with covalently bonded monolayers. *Electrochem. Solid-State Lett.* 2, 1999: 288–290.

182. Zhong, Y. L. and K. P. Loh. The chemistry of C-H bond activation on diamond. *Chem. Asian J.* 5, 2010: 1532–1540.

183. Szunerits, S. and R. Boukherroub. Different strategies for functionalization of diamond surfaces. *J. Solid State Electrochem.* 12, 2008: 1205–1218.

184. Hoffmann, R., H. Obloh, N. Tokuda, N. Yang, and C. E. Nebel. Fractional surface termination of diamond by electrochemical oxidation. *Langmuir* 28, 2012: 47–50.

185. Wildgoose, G. G., P. Abiman, and R. G. Compton. Characterising chemical functionality on carbon surfaces. *J. Mater. Chem.* 19, 2009: 4875–4886.

186. Makos, M. A., D. M. Omiatek, A. G. Ewing, and M. L. Heien. Development and characterization of a voltammetric carbon-fiber microelectrode pH sensor. *Langmuir* 26, 2010: 10386–10391.

187. Mahouche Chergui, S., N. Abbas, T. Matrab, M. Turmine, E. Bon Nguyen, R. Losno, J. Pinson, and M. M. Chehimi. Uptake of copper ions by carbon fiber/polymer hybrids prepared by tandem diazonium salt chemistry and *in situ* atom transfer radical polymerization. *Carbon* 48, 2010: 2106–2111.

188. Delamar, M., G. Desarmot, O. Fagebaume, R. Hitmi, J. Pinson, and J.-M. Savéant. Modification of carbon fiber surfaces by electrochemical reduction of aryl diazonium salts: Application to carbon epoxy composites. *Carbon* 35, 1997: 801–807.

189. Pognon, G., C. Cougnon, D. Mayilukila, and D. Bélanger. Catechol-modified activated carbon prepared by the diazonium chemistry for application as active electrode material in electrochemical capacitor. *ACS Appl. Mater. Interfaces* 4, 2012: 3788–3796.

190. Pognon, G., T. Brousse, L. Demarconnay, and D. Bélanger. Performance and stability of electrochemical capacitor based on anthraquinone modified activated carbon. *J. Power Sources* 196, 2011: 4117–4122.

191. Pognon, G., T. Brousse, and D. Bélanger. Effect of molecular grafting on the pore size distribution and the double layer capacitance of activated carbon for electrochemical double layer capacitors. *Carbon* 49, 2011: 1340–1348.

192. Weissmann, M., S. Baranton, J.-M. Clacens, and C. Coutanceau. Modification of hydrophobic/hydrophilic properties of Vulcan XC72 carbon powder by grafting of trifluoromethylphenyl and phenylsulfonic acid groups. *Carbon* 48, 2010: 2755–2764.

193. Belmont, J. A. Process for preparing carbon materials with diazonium salts and resultant carbon products. (1996) US patent 5,554,736 to Cabot Corporation (Boston, MA).

194. Paulik, M. G., P. A. Brooksby, A. D. Abell, and A. J. Downard. Grafting aryl diazonium cations to polycrystalline gold: Insights into film structure using gold oxide reduction, redox probe electrochemistry, and contact angle behavior. *J. Phys. Chem. C* 111, 2007: 7808–7815.

195. Bernard, M.-C., A. Chaussé, E. Cabet-Deliry, M. M. Chehimi, J. Pinson, F. Podvorica, and C. Vautrin-Ul. Organic layers bonded to industrial, coinage, and noble metals through electrochemical reduction of aryldiazonium salts. *Chem. Mater.* 15, 2003: 3450–3462.

196. Hapiot, P. and J. Simonet. Reduction of platinum under superdry conditions: An electrochemical approach. *Electroanal. Chem.* 23, 2010: 105–170.

197. Chaussé, A., M. M. Chehimi, N. Karsi, J. Pinson, F. Podvorica, and C. Vautrin-Ul. The electrochemical reduction of diazonium salts on iron electrodes. The formation of covalently bonded organic layers and their effect on corrosion. *Chem. Mater.* 14, 2002: 392–400.

198. Adenier, A., M.-C. Bernard, M. M. Chehimi, E. Cabet-Deliry, B. Desbat, O. Fagebaume, J. Pinson, and F. Podvorica. Covalent modification of iron surfaces by electrochemical reduction of aryldiazonium salts. *J. Am. Chem. Soc.* 123, 2001: 4541–4549.

199. Kullapere, M., L. Matisen, A. Saar, V. Sammelselg, and K. Tammeveski. Electrochemical behaviour of nickel electrodes modified with nitrophenyl groups. *Electrochem. Comm.* 9, 2007: 2412–2417.

200. Devillers, S., B. Barthélémy, I. Fery, J. Delhalle, and Z. Mekhalif. Functionalization of nitinol surface toward a versatile platform for post-grafting chemical reactions. *Electrochim. Acta* 56, 2011: 8129–8137.
201. Cottineau, T., M. Morin, and D. Bélanger. Surface band structure of aryl-diazonium modified p-Si electrodes determined by X-ray photoelectron spectroscopy and electrochemical measurements. *RSC Adv.* 3, 2013: 23649–23657.
202. El Hadja, F., A. Amiar, M. Cherkaoui, J.-N. Chazalviel, and F. Ozanam. Study of organic grafting of the silicon surface from 4-nitrobenzene diazonium tetrafluoroborate. *Electrochim. Acta* 70, 2012: 318–324.
203. Chen, M., K. Kobashi, B. Chen, M. Lu, and J. M. Tour. Functionalized self-assembled InAs/GaAs quantum-dot structures hybridized with organic molecule. *Adv. Funct. Mater.* 20, 2010: 469–475.
204. Chiou, B.-H., Y.-T. Tsai, and C. M. Wang. Phenothiazine-modified electrodes: A useful platform for protein adsorption study. *Langmuir* 30, 2014: 1550–1556.
205. Chung, D.-J., S.-H. Oh, S. Komathi, A. I. Gopalan, K.-P. Lee, and S.-H. Choi. One-step modification of various electrode surfaces using diazonium salt compounds and the application of this technology to electrochemical DNA (E-DNA) sensors. *Electrochim. Acta* 76, 2012: 394–403.
206. Maldonado, S., T. J. Smith, R. D. Williams, S. Morin, E. Barton, and K. J. Stevenson. Surface modification of indium tin oxide via electrochemical reduction of aryldiazonium cations. *Langmuir* 22, 2006: 2884–2891.
207. Lamberti, F., S. Agnoli, L. Brigo, G. Granozzi, M. Giomo, and N. Elvassore. Surface functionalization of fluorine-doped tin oxide samples through electrochemical grafting. *ACS Appl. Mater. Interfaces* 5, 2013: 12887–12894.
208. Merson, A., Th. Dittrich, Y. Zidon, J. Rappich, and Y. Shapira. Charge transfer from TiO_2 into adsorbed benzene diazonium compounds. *Appl. Phys. Lett.* 85, 2004: 1075–1076.
209. Li, F., Y. Feng, P. Dong, L. Yang, and B. Tang. Gold nanoparticles modified electrode via simple electrografting of *in situ* generated mercaptophenyl diazonium cations for development of DNA electrochemical biosensor. *Biosens. Bioelec.* 26, 2011: 1947–1952.
210. Orefuwa, S. A., M. Ravanbakhsh, S. N. Neal, J. B. King, and A. A. Mohamed. Robust organometallic gold nanoparticles. *Organometallics* 33, 2014: 439–442.
211. Ktari, N., J. Quinson, B. Teste, J.-M. Siaugue, F. Kanoufi, and C. Combellas. Immobilization of magnetic nanoparticles onto conductive surface modified by diazonium chemistry. *Langmuir* 28, 2012: 12671–12680.
212. Kainz, Q. M., A. Schätz, A. Zöpfl, W. J. Stark, and O. Reiser. Combined covalent and noncovalent functionalization of nanomagnetic carbon surfaces with dendrimers and BODIPY fluorescent dye. *Chem. Mater.* 23, 2011: 3606–3613.
213. Griffete, N., F. Herbst, J. Pinson, S. Ammar, and C. Mangeney. Preparation of water-soluble magnetic nanocrystals using aryl diazonium salt chemistry. *J. Am. Chem. Soc.* 133, 2011: 1646–1649.
214. Szunerits, S. and R. Boukherroub. Diamond nanoparticles. A review of selected surface modification. Methods for bioconjugation, biology and medicine, in: *Applied Surface Chemistry of Nanomaterials*, M. M. Chehimi, J. Pinson (Eds.), Nova Science Publishers Inc., New York, 2013, pp. 3–32.
215. Yang, N., W. Smirnov, and W. E. Nebel. Diamond nanostructure technologies, properties and electrochemical applications, in: *Applied Surface Chemistry of Nanomaterials*, M.M. Chehimi, J. Pinson (Eds.), Nova Science Publishers Inc., New York, 2013, pp. 33–54.
216. Salmi, Z., A. Lamouri, P. Decorse, M. Jouini, A. Boussadi, J. Achard, A. Gicquel, S. Mahouche-Chergui, B. Carbonnier, and M. M. Chehimi. Grafting polymer–protein bioconjugate to boron-doped diamond using aryldiazonium coupling agents. *Diamond Relat. Mater.* 40, 2013: 60–68.

217. Yeap, W. S., S. Chen, and K. P. Loh. Detonation nanodiamond: An organic platform for the Suzuki coupling of organic molecules. *Langmuir* 25, 2009: 185–191.

218. Chen, B., M. Lu, A. K. Flatt, F. Maya, and J. M. Tour. Chemical reactions in monolayer aromatic films on silicon surfaces. *Chem. Mater.* 20, 2008: 61–64.

219. Peng, X. and S. S. Wong. Functional covalent chemistry of carbon nanotube surfaces. *Adv. Mater.* 21, 2009: 625–642.

220. Singh, P., S. Campidelli, S. Giordani, D. Bonifazi, A. Bianco, and M. Prato. Organic functionalisation and characterisation of single-walled carbon nanotubes. *Chem. Soc. Rev.* 38, 2009: 2214–2230.

221. Lu, F., L. Gu, M. J. Meziani, X. Wang, P. G. Luo, L. M. Veca, L. Cao, and Y.-P. Sun. Advances in bioapplications of carbon nanotubes. *Adv. Mater.* 21, 2009: 139–152.

222. Balasubramanian, K. and M. Burghard. Electrochemically functionalized carbon nanotubes for device applications. *J. Mater. Chem.* 18, 2008: 3071–3083.

223. Guldi, D. M., G. M. Aminur Rahman, V. Sgobba, and C. Ehli. Multifunctional molecular carbon materials—from fullerenes to carbon nanotubes. *Chem. Soc. Rev.* 35, 2006: 471–487.

224. Bahr, J. L., J. Yang, D. V. Kosynkin, M. J. Bronikowski, R. E. Smalley, and J. M. Tour. Functionalization of carbon nanotubes by electrochemical reduction of aryl diazonium salts: A bucky paper electrode. *J. Am. Chem. Soc.* 123, 2001: 6536–6542.

225. Marcoux, P. R., P. Hapiot, P. Batail, and J. Pinson. Electrochemical functionalization of nanotube films: Growth of aryl chains on single-walled carbon nanotubes. *New. J. Chem.* 28, 2004: 302–307.

226. Kooi, S. E., U. Schlecht, M. Burghard, and K. Kern. Electrochemical modification of single carbon nanotubes. *Angew. Chem. Int. Ed.* 41, 2002: 1353–1355.

227. Gohier, A., F. Nekelson, M. Helezen, P. Jégou, G. Deniau, S. Palacin, and M. Mayne-L'Hermite. Tunable grafting of functional polymers onto carbon nanotubes using diazonium chemistry in aqueous media. *J. Mater. Chem.* 21, 2011: 4615–4622.

228. Bahr, J. L. and J. M. Tour. Highly functionalized carbon nanotubes using *in situ* generated diazonium compounds. *Chem. Mater.* 13, 2001: 3823–3824.

229. Salice, P., E. Fabris, C. Sartorio, D. Fenaroli, V. Figa, M. P. Casaletto, S. Cataldo, B. Pignataro, and E. Menna. An insight into the functionalisation of carbon nanotubes by diazonium chemistry: Towards a controlled decoration. *Carbon* 74, 2014: 73–82.

230. Englert, J. M., C. Dotzer, G. Yang, M. Schmid, C. Papp, J. M. Gottfried, H.-P. Steinrück, E. Spiecker, F. Hauke, and A. Hirsch. Covalent bulk functionalization of graphene. *Nature Chem.* 3, 2011: 279–286.

231. Park, J. and M. Yan. Covalent functionalization of graphene with reactive intermediates. *Acc. Chem. Res.* 46, 2013: 181–189.

232. Koehler, F. M. and W. J. Stark. Organic synthesis on graphene. *Acc. Chem. Res.* 46, 2013: 2297–2306.

233. Dustin, K. J. and J. M. Tour. Graphene: Powder, flakes, ribbons, and sheets. *Acc. Chem. Res.* 46, 2013: 2307–2318.

234. Bekyarova, E., S. Sarkar, F. Wang, M. E. Itkis, I. Kalinina, X. Tian, and R. C. Haddon. Effect of covalent chemistry on the electronic structure and properties of carbon nanotubes and graphene. *Acc. Chem. Res.* 46, 2013: 65–76.

235. Azevedo, J., L. Fillaud, C. Bourdillon, J.-M. Noël, F. Kanoufi, B. Jousselme, V. Derycke, S. Campidelli, and R. Cornut. Localized reduction of graphene oxide by electrogenerated naphthalene radical anions and subsequent diazonium electrografting. *J. Am. Chem. Soc.* 136, 2014: 4833–4836.

236. Chehimi, M. M., J. Pinson, Z. Salmi. Carbon nanotubes: Surface modification and applications, in: *Applied Chemistry of Nanomaterials,* M. M. Chehimi, J. Pinson (Eds.), Nova Science Publishers Inc., New York, 2013, pp. 57–94.

237. Do, Y.-J., J.-H. Lee, H. Choi, J.-H. Han, C.-H. Chung, M.-G. Jeong, M. S. Strano, and W.-J. Kim. Manipulating electron transfer between single-walled carbon nanotubes and diazonium salts for high purity separation by electronic type. *Chem. Mater* 24, 2012: 4146–4151.

238. Ghosh, S. and C. N. R. Rao. Separation of metallic and semiconducting single-walled carbon nanotubes through fluorous chemistry. *Nano Res.* 2, 2009: 183–191.

239. Sharma, R., J. H. Baik, C. J. Perera, and M. S. Strano. Anomalously large reactivity of single graphene layers and edges toward electron transfer chemistries. *Nano Lett.* 10, 2010: 398–405.

240. Li, Q., C. Batchelor-McAuley, N. S. Lawrence, R. S. Hartshorne, and R. G. Compton. The synthesis and characterization of controlled thin sub-monolayer films of 2-anthraquinonyl groups on graphite surfaces. *New J. Chem.* 35, 2011: 2462–2470.

241. Nowak, A. M. and R. L. McCreery. Characterization of carbon/nitroazobenzene/titanium molecular electronic junctions with photoelectron and Raman spectroscopy. *Anal. Chem.* 76, 2004: 1089–1097.

242. Anariba, F., U. Viswanathan, D. F. Bocian, and R. L. McCreery. Determination of the structure and orientation of organic molecules tethered to flat graphitic carbon by ATR-FT-IR and Raman spectroscopy. *Anal. Chem.* 78, 2006: 3104–3112.

243. Lu, M., T. He, and J. M. Tour. Surface grafting of ferrocene-containing triazene derivatives on Si(100). *Chem. Mater.* 20, 2008: 7352–7355.

244. Leroux, Y. R., H. Fei, J.-M. Noël, C. Roux, and P. Hapiot. Efficient covalent modification of a carbon surface: Use of a silyl protecting group to form an active monolayer. *J. Am. Chem. Soc.* 132, 2010: 14039–14041.

245. Leroux, Y. R. and P. Hapiot. Nanostructured monolayers on carbon substrates prepared by electrografting of protected aryldiazonium salts. *Chem. Mater.* 25, 2013: 489–495.

246. Lee, L., P. A. Brooksby, Y. R. Leroux, P. Hapiot, and A. J. Downard. Mixed monolayer organic films via sequential electrografting from aryldiazonium ion and arylhydrazine solutions. *Langmuir* 29, 2013: 3133–3139.

247. Nielsen, L. T., K. H. Vase, M. Dong, F. Besenbacher, S. U. Pedersen, and K. Daasbjerg. Electrochemical approach for constructing a monolayer of thiophenolates from grafted multilayers of diaryl disulfides. *J. Am. Chem. Soc.* 129, 2007: 1888–1889.

248. Malmos, K., M. Dong, S. Pillai, P. Kingshott, F. Besenbacher, S. U. Pedersen, and K. Daasbjerg. Using a hydrazone-protected benzenediazonium salt to introduce a near-monolayer of benzaldehyde on glassy carbon surfaces. *J. Am. Chem. Soc.* 131, 2009: 4928–4936.

249. Lee, L., H. Ma, P. A. Brooksby, S. A. Brown, Y. R. Leroux, P. Hapiot, and A. J. Downard. Covalently anchored carboxyphenyl monolayer via aryldiazonium ion grafting: A well-defined reactive tether layer for on-surface chemistry. *Langmuir* 30, 2014: 7104–7111.

250. Menanteau, T., E. Levillain, and T. Breton. Electrografting via diazonium chemistry: From multilayer to monolayer using radical scavenger. *Chem. Mater.* 25, 2013: 2905–2909.

251. Brooksby, P. A. and A. J. Downard. Electrochemical and atomic force microscopy study of carbon surface modification via diazonium reduction in aqueous and acetonitrile solutions. *Langmuir* 20, 2004: 5038–5045.

252. Allongue, P., C. Henry de Villeneuve, G. Cherouvrier, R. Cortès, and M.-C. Bernard. Phenyl layers on H-Si(111) by electrochemical reduction of diazonium salts: Monolayer versus multilayer formation. *J. Electroanal. Chem.* 550–551, 2003: 161–174.

253. Uetsuka, H., D. Shin, N. Tokuda, K. Saeki, and C. E. Nebel. Electrochemical grafting of boron-doped single-crystalline chemical vapor deposition diamond with nitrophenyl molecules. *Langmuir* 23, 2007: 3466–3472.

254. Yang, N., H. Zhuang, R. Hoffmann, W. Smirnov, J. Hees, X. Jiang, and C. E. Nebel. Electrochemistry of nanocrystalline 3C silicon carbide films. *Chem. Eur. J.* 18, 2012: 6514–6519.

255. Yu, S. S. C., E. S. Q. Tan, R. T. Jane, and A. J. Downard. An electrochemical and XPS study of reduction of nitrophenyl films covalently grafted to planar carbon surfaces. *Langmuir* 23, 2007: 11074–11082.

256. Brooksby, P. A. and A. J. Downard. Multilayer nitroazobenzene films covalently attached to carbon. An AFM and electrochemical study. *J. Phys. Chem. B* 109, 2005: 8791–8798.

257. Lehr, J., B. E. Williamson, B. S. Flavel, and A. J. Downard. Reaction of gold substrates with diazonium salts in acidic solution at open-circuit potential. *Langmuir* 25, 2009: 13503–13509.

258. Ceccato, M., L. T. Nielsen, J. Iruthayaraj, M. Hinge, S. U. Pedersen, and K. Daasbjerg. Nitrophenyl groups, in diazonium-generated multilayered films: Which are electrochemically responsive? *Langmuir* 26, 2010, pp. 10812–10821.

259. Haccoun, J., C. Vautrin-Ul, A. Chaussé, A. Adenier. Electrochemical grafting of organic coating onto gold surfaces: Influence of the electrochemical conditions on the grafting of nitrobenzene diazonium salt. *Progress Org. Coatings* 63, 2008: 18–24.

260. Anariba, F., S. H. DuVall, and R. L. McCreery. Mono- and multilayer formation by diazonium reduction on carbon surfaces monitored with atomic force microscopy "scratching." *Anal. Chem.* 75, 2003: 3837–3844.

261. Munteanu, S., N. Garraud, J. P. Roger, F. Amiot, J. Shi, Y. Chen, C. Combellas, and F. Kanoufi. *In situ*, real time monitoring of surface transformation: Ellipsometric microscopy imaging of electrografting at microstructured gold surfaces. *Anal. Chem.* 85, 2013: 1965–1971.

262. Diget, J. S., H. N. Petersen, H. Schaarup-Jensen, A. U. Sørensen, T. T. Nielsen, S. U. Pedersen, K. Daasbjerg, K. L. Larsen, and M. Hinge. Synthesis of β-cyclodextrin diazonium salts and electrochemical immobilization onto glassy carbon and gold surfaces. *Langmuir* 28, 2012: 16828–16833.

263. Kullapere, M., F. Mirkhalaf, and K. Tammeveski. Electrochemical behaviour of glassy carbon electrodes modified with aryl groups. *Electrochim. Acta* 56, 2010: 166–173.

264. Downard, A. J. and M. J. Prince. Barrier properties of organic monolayers on glassy carbon electrodes. *Langmuir* 17, 2001: 5581–5586.

265. Gui, A. L., G. Liu, M. Chockalingam, G. Le Saux, J. B. Harper, and J. J. Gooding. A comparative study of modifying gold and carbon electrode with 4-sulfophenyl diazonium salt. *Electroanalysis* 22, 2010: 1283–1289.

266. Noël, J.-M., B. Sjöberg, R. Marsac, D. Zigah, J.-F. Bergamini, A. Wang, S. Rigaut, P. Hapiot, and C. Lagrost. Flexible strategy for immobilizing redox-active compounds using in situ generation of diazonium salts. Investigations of the blocking and catalytic properties of the layers. *Langmuir* 25, 2009: 12742–12749.

267. Hauquier, F., T. Matrab, F. Kanoufi, and C. Combellas. Local direct and indirect reduction of electrografted aryldiazonium/gold surfaces for polymer brushes patterning. *Electrochim. Acta* 54, 2009: 5127–5136.

268. Latus, A., J.-M. Noël, E. Volanschi, C. Lagrost, and P. Hapiot. Scanning electrochemical microscopy studies of glutathione-modified surfaces. An erasable and sensitive-to-reactive oxygen species surface. *Langmuir* 27, 2011: 11206–11211.

269. Fave, C., V. Noël, J. Ghilane, G. Trippé-Allard, H. Randriamahazaka, and J.-C. Lacroix. Electrochemical switches based on ultrathin organic films: From diode-like behavior to charge transfer transparency. *J. Phys. Chem.* 112, 2008: 18638–18643.

270. Stockhausen, V., J. Ghilane, P. Martin, G. Trippé-Allard, H. Randriamahazaka, and J.-C. Lacroix. Grafting oligothiophenes on surfaces by diazonium electroreduction: A step toward ultrathin junction with well-defined metal/oligomer interface. *J. Am. Chem. Soc.* 131, 2009: 14920–14927.

271. Polsky, R., J. C. Harper, S. M. Dirk, D. C. Arango, D. R. Wheeler, and S. M. Brozik. Diazonium-functionalized horseradish peroxidase immobilized via addressable electrodeposition: Direct electron transfer and electrochemical detection. *Langmuir* 23, 2007: 364–366.

272. Garretta, D. J., P. Jenkins, M. I. J. Polson, D. Leech, K. H. R. Baron, and A. J. Downard. Diazonium salt derivatives of osmium bipyridine complexes: Electrochemical grafting and characterization of modified surfaces. *Electrochim. Acta* 56, 2011: 2213–2220.

273. Yan, H., A. J. Bergren, R. L. McCreery, M. L. Della Rocca, P. Martin, P. Lafarge, and J. C. Lacroix. Activationless charge transport across 4.5 to 22 nm in molecular electronic junctions. *Proc. Natl. Acad. Sci. USA* 110, 2013: 5326–5330.

274. McCreery, R. L. Electron transport and redox reactions in molecular electronic junctions. *Chem. Phys. Chem.* 10, 2009: 2387–2391.

275. Rabache, V., J. Chaste, P. Petit, M. L. Della Rocca, P. Martin, J. C. Lacroix, R. L. McCreery, and P. Lafarge. Direct observation of large quantum interference effect in anthraquinone solid-state junctions. *J. Am. Chem. Soc.* 135, 2013: 10218–10221.

276. Chernyy, S., K. Torbensen, J. Iruthayaraj, S. U. Pedersen, and K. Daasbjerg. Elucidation of the mechanism of redox grafting of diazotated anthraquinone. *Langmuir* 28, 2012: 9573–9582.

277. A. Bousquet, M. Ceccato, M. Hinge, S. U. Pedersen, and K. Daasbjerg. Redox grafting of diazotated anthraquinone as a means of forming thick conducting organic films. *Langmuir* 28, 2012: 1267–1275.

278. Louault, C., M. D'Amours, and D. Bélanger. The electrochemical grafting of a mixture of substituted phenyl groups at a glassy carbon electrode surface. *Chem. Phys. Chem.* 9, 2008: 1164–1170.

279. Gui, A. L., H. M. Yau, D. S. Thomas, M. Chockalingam, J. B. Harper, and J. J. Gooding. Using supramolecular binding motifs to provide precise control over the ratio and distribution of species in multiple component films grafted on surfaces: Demonstration using electrochemical assembly from aryl diazonium salts. *Langmuir* 29, 2013: 4772–4781.

280. Santos, L., J. Ghilane, J. C. Lacroix. Formation of mixed organic layers by stepwise electrochemical reduction of diazonium compounds. *J. Am. Chem. Soc.* 134, 2012: 5476–5479.

281. Usrey, M. L., E. S. Lippmann, and M. S. Strano. Evidence for a two-step mechanism in electronically selective single-walled carbon nanotube reactions. *J. Am. Chem. Soc.* 127, 2005: 16129–16135.

282. Daasbjerg, K. and K. J. Sehested. Reduction of substituted benzenediazonium salts by solvated electrons in aqueous neutral solution studied by pulse radiolysis. *J. Phys. Chem. A* 106, 2002: 11098–1106.

283. Daasbjerg, K. and K. J. Sehested. Reduction of substituted benzenediazonium salts by hydrogen atoms in aqueous acidic solution studied by pulse radiolysis. *J. Phys. Chem. A* 107, 2003: 4462–4469.

284. Suehiro, T., S. Masuda, T. Tashiro, R. Nakausa, M. Taguchi, A. Koike, and A. Rieker. Formation and identification of aryldiazenyl radicals using esr technique. *Bull. Chem. Soc. Jpn.* 59, 1986: 1877–1886.

285. Suehiro, T., S. Masuda, T. R. Nakausa, M. Taguchi, A. Mori, A. Koike, and M. Date. Decay reactions of aryldiazenyl radicals in solution. *Bull. Chem. Soc. Jpn.* 60, 1987: 3221–3330.

286. Kullapere, M., M. Marandi, V. Sammelselg, H. A. Menezes, G. Maia, K. Tammeveski. Surface modification of gold electrodes with anthraquinone diazonium cations. *Electrochem. Comm.* 11, 2009: 405–408.

287. Wetzel, A., G. Pratsch, R. Kolb, and M. R. Heinrich. Radical arylation of phenols, phenyl ethers, and furans. *Chem. Eur. J.* 16, 2010: 2547–2556.

288. Barbier, B., J. Pinson, G. Desarmot, M. Sanchez. Electrochemical bonding of amines to carbon fiber surfaces toward improved carbon-epoxy composites. *J. Electrochem. Soc.* 137, 1990: 1757–1764.

289. Gam-Derouich, S., S. Mahouche-Chergui, M. M. Chehimi, and H. Ben Romdhane. Polymer grafting to aryl diazonium-modified materials: Methods and applications, in: *Aryl Diazonium Salts: New Coupling Agents in Polymer and Surface Science,* M. M. Chehimi, (Ed.), 2012 Wiley Chap. 6, pp. 125–157.

290. Wang, Y., L. Meng, L. Fan, L. Ma, M. Qi, J. Yu, and Y. Huang. Enhanced interfacial properties of carbon fiber composites via aryldiazonium reaction "on water." *Appl. Surf. Sci.* 316, 2014: 366–372.

291. Adenier, A, E. Cabet-Deliry, T. Lalot, J. Pinson, and F. Podvorica. Attachment of polymers to organic moieties covalently bonded to iron surfaces. *Chem. Mater.* 14, 2002: 4576–4585.

292. Li, H., F. Cheng, A. M. Duft, A. Adronov. Functionalization of single-walled carbon nanotubes with well-defined polystyrene by "Click" coupling. *J. Am. Chem. Soc.* 127, 2005: 14518–14524.

293. Mahouche Chergui, S., A. Ledebt, F. Mammeri, F. Herbst, B. Carbonnier, H. Ben Romdhane, M. Delamar, M. M. Chehimi. Hairy carbon nanotube@nano-Pd heterostructures: Design, characterization, and application in Suzuki C-C coupling reaction. *Langmuir* 26, 2010: 16115–16121.

294. Salmi, Z., C. Epape, S. Mahouche-Chergui, B. Carbonnier, M. Omastová, M. M. Chehimi. Multiwalled carbon nanotube-clicked poly(4-vinyl pyridine) as a hairy platform for the immobilization of gold nanoparticles. *J. Colloid Sci. Biotechnol.* 2, 2013: 53–61.

295. Mahouche, S., N. Mekni, L. Abbassi, P. Lang, C. Perruchot, M. Jouini, F. Mammeri, M. Turmine, H. Ben Romdhane, and M. M. Chehimi. Tandem diazonium salt electroreduction and click chemistry as a novel, efficient route for grafting macromolecules to gold surface *Surf. Sci.* 603, 2009: 3205–3211.

296. Li, H. and A. Adronov. Water-soluble SWCNTs from sulfonation of nanotube-bound polystyrene. *Carbon*, 45, 2007: 984–990.

297. Ratheesh Kumar, V. K. and K. R. Gopidas. Synthesis and characterization of goldnanoparticle-cored dendrimers stabilized by metal–carbon bonds. *Chem. Asian. J.* 5, 2010: 887–896.

298. Ratheesh Kumar, V. K. and K. R. Gopidas. Palladium nanoparticle-cored G1-dendrimer stabilized by carbon-Pd bonds: Synthesis, characterization and use as chemoselective, room temperature hydrogenation catalyst. *Tetrahedron Lett.* 52, 2011: 3102–3105.

299. Ratheesh Kumar, V. K., S. Krishnakumar, and K. R. Gopidas. Synthesis, characterization and catalytic applications of palladium nanoparticle-cored dendrimers stabilized by metal–carbon bonds. *Eur. J. Org. Chem.* 2012, 2012: 3447–3458.

300. Xiong, Z., T. Gu, and X. Wang. Self-assembled multilayer films of sulfonated graphene and polystyrene-based diazonium salt as photo-cross-linkable supercapacitor electrodes. *Langmuir* 30, 2014: 522–532.

301. Wang, G. J., S. Z. Huang, Y. Wang, L. L. Liu, J. Qiu, and Y. Li. Synthesis of water-soluble single-walled carbon nanotubes by RAFT polymerization. *Polymer* 48, 2007: 728–733.

302. Liu J., L. Cui, N. Kong, C. J. Barrow, and W. Yang. RAFT controlled synthesis of graphene/polymer hydrogel with enhanced mechanical property for pH-controlled drug release. *Eur. Polym. J.* 50, 2014: 9–17.

303. Griffete, N., H. Li, A. Lamouri, C. Redeuilh, K. Chen, C. Z. Dong, S. Nowak, S. Ammar, and C. Mangeney. Magnetic nanocrystals coated by molecularly imprinted polymers for the recognition of bisphenol A. *J. Mater. Chem.* 22, 2012: 1807–1811.

304. Ahmad, R., N. Griffete, A. Lamouri, and C. Mangeney. Functionalization of magnetic nanocrystals by oligo (ethylene oxide) chains carrying diazonium and iniferter end groups. *J. Colloid Interface Sci.* 407, 2013: 210–214.

305. Stephenson, J. J., A. K. Sadana, A. L. Higginbotham, and J. M. Tour. Stephenson, J. J., A. K. Sadana, A. L. Higginbotham, and J. M. Tour. Highly functionalized and soluble multiwalled carbon nanotubes by reductive alkylation and arylation: The Billups reaction. *Chem. Mater.* 18, 2006: 4658–4661.

306. M. Santos, J. Ghilane, C. Fave, P. C. Lacaze, H. Randriamahazaka, L. M. Abrantes, and J. C. Lacroix. Electrografting polyaniline on carbon through the electroreduction of diazonium salts and the electrochemical polymerization of aniline. *J. Phys. Chem. C* 112, 2008: 16103–16109.

307. Lacroix, J.-C., J. Ghilane, L. Santos, G. Trippé-Allard, P. Martin, and H. Randriamahazaka. Electrografting of conductive oligomers and polymers, in: *Aryl Diazonium Salts. New Coupling Agents in Polymer and Surface Science*, M.M. Chehimi, Wiley-VCH, Weinheim, 2012, Chap. 8, pp. 181–195.

308. Blacha, A., P. Koscielniak, M. Sitarz, J. Szuber, and J. Zak. Pedot brushes electrochemically synthesized on thienyl-modified glassy carbon surfaces. *Electrochim. Acta* 62, 2012: 441–446.

309. Mekki, A., S. Samanta, A. Singh, Z. Salmi, R. Mahmoud, M. M. Chehimi, and D. K. Aswal. Core/shell, protuberance-free multiwalled carbon nanotube/polyaniline nanocomposites via interfacial chemistry of aryl diazonium salts. *J. Colloid Interface Sci.* 418, 2014: 185–192.

310. Kato, M., M. Kamigaito, M. Sawamoto, and T. Higashimura. Polymerization of methyl methacrylate with the carbon tetrachloride/dichlorotris-(triphenylphosphine) ruthenium(II)/methylaluminum bis(2,6-di-tert-butylphenoxide) initiating system: Possibility of living radical polymerization. *Macromolecules* 28, 1995: 1721–1723.

311. Wang, J. S. and K. Matyjaszewski. Controlled living radical polymerization—Atom-transfer radical polymerization in the presence of transition-metal complexes. *J. Am. Chem. Soc.* 117, 1995: 5614–5615.

312. Matyjaszewski, K., P. J. Miller, N. Shukla, B. Immaraporn, A. Gelman, B. B. Luokala, T. M. Siclovan, G. Kickelbick, T. Vallant, H. Hoffmann, and T. Pakula. Polymers at interfaces: Using atom transfer radical polymerization in the controlled growth of homopolymers and block copolymers from silicon surfaces in the absence of untethered sacrificial initiator. *Macromolecules* 32, 1999: 8716–8724.

313. Edmondson, S., V. L. Osborne, W. T. S. Huck. Polymer brushes via surface-initiated polymerizations. *Chem. Soc. Rev.* 33, 2004: 14–22.

314. Barbey, R., L. Lavanant, D. Paripovic, N. Schüwer, C. Sugnaux, S. Tugulu, and H.-A. Klok. Polymer brushes via surface-initiated controlled radical polymerization: Synthesis, characterization, properties, and applications. *Chem. Rev.* 109, 2009: 5437–5527.

315. Matrab, T., M. M. Chehimi, C. Perruchot, A. Adenier, V. Guillez, M. Save, B. Charleux, E. Cabet-Delry, and J. Pinson. Novel approach for metallic surface-initiated atom transfer radical polymerization using electrografted initiators based on aryl diazonium salts. *Langmuir* 21, 2005: 4686–4694.

316. Liu, T., R. Casado-Portilla, J. Belmont, and K. Matyjaszewski. ATRP of butyl acrylates from functionalized carbon black surfaces. *J. Polym. Sci. A Polym. Chem.* 43, 2005: 4695–4709.

317. Iruthayaraj, J., S. Chernyy, M. Lillethorup, M. Ceccato, T. Røn, M. Hinge, P. Kingshott, F. Besenbacher, S. U. Pedersen, and K. Daasbjerg. On surface-initiated atom transfer radical polymerization using diazonium chemistry to introduce the initiator layer. *Langmuir* 27, 2011: 1070–1078.

318. Matrab, T., M. Save, B. Charleux, J. Pinson, E. Cabet-Delry, A. Adenier, M. M. Chehimi, and M. Delamar. Grafting densely-packed poly(n-butyl methacrylate) chains from an iron substrate by aryl diazonium surface-initiated ATRP: XPS monitoring. *Surf. Sci.* 601, 2007: 2357–2366.

319. Matrab, T., M. M. Chehimi, J. Pinson, S. Slomkowski, and T. Basinska. Growth of polymer brushes by atom transfer radical polymerization on glassy carbon modified by electro-grafted initiators based on aryl diazonium salts. *Surf. Interface Anal.* 38, 2006: 565–568.

320. Nguyen, M. N., T. Matrab, C. Badre, M. Turmine, and M. M. Chehimi. Interfacial aspects of polymer brushes prepared on conductive substrates by aryl diazonium salt surface-initiated ATRP *Surf. Interface Anal.* 40, 2008: 412–417.

321. Lillethorup, M., K. Shimizu, N. Plumeré, S. U. Pedersen, K. Daasbjerg. Surface-attached poly(glycidyl methacrylate) as a versatile platform for creating dual-functional polymer brushes. *Macromolecules* 47, 2014: 5081–5088.

322. Lillethorup, M. K. Torbensen, M. Ceccato, S. U. Pedersen, and K. Daasbjerg. Electron transport through a diazonium-based initiator layer to covalently attached polymer brushes of ferrocenylmethyl methacrylate. *Langmuir* 29, 2013: 13595–13604.

323. Jin, G.-P., Y. Fu, X.-C. Bao, X.-S. Feng, Y. Wang, and W.-H. Liu. Electrochemically mediated atom transfer radical polymerization of iminodiacetic acid-functionalized poly(glycidyl methacrylate)grafted at carbon fibers for nano-nickel recovery from spent electroless nickel plating baths. *J. Appl. Electrochem.* 44, 2014: 621–629.

324. Matrab, T., J. Chancolon, M. Mayne-L'Hermite, J. N. Rouzaud, G. Deniau, J.-P. Boudou, M. M. Chehimi, and M. Delamar. Atom transfer radical polymerization (ATRP) initiated by aryl diazonium salts: A new route for surface modification of multiwalled carbon nanotubes by tethered polymer chains. *Colloids Surf. A Physicochem. Asp.* 287, 2006: 217–221.

325. Wu, W., N. V. Tsarevsky, J. L. Hudson, J. M. Tour, K. Matyjaszewski, and T. Kowalewski. "Hairy" single-walled carbon nanotubes prepared by atom transfer radical polymerization. *Small* 3, 2007: 1803–1810.

326. Fang, M., K. Wang, H. Lu, Y. Yang, S. Nutt. Covalent polymer functionalization of graphene nanosheets and mechanical properties of composites. *J. Mater. Chem.* 19, 2009: 7098–7105.

327. Gehan H., L. Fillaud, M. M. Chehimi, J. Aubard, A. Hohenau, N. Felidj, C. Mangeney. Thermo-induced electromagnetic coupling in gold/polymer hybrid plasmonic structures probed by surface-enhanced Raman scattering. *ACS Nano* 4, 2010: 6491–6500.

328. Chernyy, S., J. Iruthayaraj, M. Ceccato, M. Hinge, S. Uttrup Pedersen, and K. Daasbjerg. Elucidation of the mechanism of surface-initiated atom transfer radical polymerization from a diazonium-based initiator layer. *J. Polym. Sci. A: Polym. Chem. Ed.* 50, 2012: 4465–4475.

329. Shimizu, K., K. Malmos, A. H. Holm, S. U. Pedersen, K. Daasbjerg, and M. Hinge. Improved adhesion between PMMA and stainless steel modified with PMMA brushes. *ACS Appl. Mater. Interfaces* 6, 2014: 21308–21315.

330. Shaulov Y., R. Okner, Y. Levi, N. Tal, V. Gutkin, D. Mandler, and A. J. Domb. Poly(methyl methacrylate) grafting onto stainless steel surfaces: Application to drug-eluting stents. *ACS Appl. Mater. Interfaces* 1, 2009: 2519–2528.

331. Matrab, T., M. M. Chehimi, J. P. Boudou, F. Benedic, J. Wang, N. N. Naguib, and J. A. Carlisle. Surface functionalization of ultrananocrystalline diamond using atom transfer radical polymerization (ATRP) initiated by electro-grafted aryldiazonium salts. *Diamond Relat. Mater.* 15, 2006: 639–644.

332. Dahoumane, S. A., M. N. Nguyen, A. Thorel, J. P. Boudou, M. M. Chehimi, and C. Mangeney. Protein-functionalized hairy diamond nanoparticles. *Langmuir* 25, 2009: 9633–9638.

333. Gam-Derouich, S., B. Carbonnier, M. Turmine, P. Lang, M. Jouini, D. Ben Hassen-Chehimi, and M M. Chehimi. Electrografted aryl diazonium initiators for surface-confined photopolymerization: A new approach to designing functional polymer coatings. *Langmuir* 26, 2010: 11830–11840.

334. Gam-Derouich, S., S. Mahouche-Chergui, M. Turmine, J.-Y. J.-Y. Piquemal, D. Ben Hassen-Chehimi, M. Omastová, and M. M. Chehimi. A versatile route for surface modification of carbon, metals and semi-conductors by diazonium salt-initiated photo-polymerization. *Surf. Sci.* 605, 2011: 1889–1899.

335. Gam-Derouich, S., A. Lamouri, C. Redeuilh, P. Decorse, F. Maurel, B. Carbonnier, S. Beyazıt, G. Yilmaz, Y. Yagci, and M. M. Chehimi. Diazonium salt-derived 4-(dimethyl-amino)phenyl groups as hydrogen donors in surface-confined radical photopolymerization for bioactive poly(2-hydroxyethyl methacrylate) grafts. *Langmuir* 28, 2012: 8035–8045.

336. Haupt, K., A. V. Linares, M. Bompart, and B. Tse Sum Bui. Micro and nanofabrication of molecularly imprinted polymers. *Top. Curr. Chem.* 325, 2012: 83–110.

337. Fuchs, Y., O. Soppera, and K. Haupt. Photopolymerization and photostructuring of molecularly imprinted polymers for sensor applications- A review. *Anal. Chim. Acta* 717, 2012: 7–20.

338. C. Malitesta, E. Mazzotta, R. A. Picca, A. Poma, I. Chianella, S. A. Piletsky, MIP sensors—the electrochemical approach. *Anal. Bioanal. Chem.* 402, 2012: 1827–1846.

339. Bakas, I., Z. Salmi, S. Gam-Derouich, M. Jouini, S. Lépinay, B. Carbonnier, A. Khlifi, R. Kalfat, F. Geneste, Y. Yagci, and M. M. Chehimi. Molecularly imprinted polymeric sensings layers grafted from aryl diazonium-modified surfaces for electroanalytical applications. A mini review. *Surf. Interface Anal.* 46, 2014: 1014–1020.

340. Gam-Derouich, S., M. N. Nguyen, A. Madani, N. Maouche, P. Lang, C. Perruchot, and M. M. Chehimi. Aryl diazonium salt surface chemistry and ATRP for the preparation of molecularly imprinted polymer grafts on gold substrates. *Surf. Interface Anal.* 42, 2010: 1050–1056.

341. Khlifi, A., S. Gam-Derouich, M. Jouini, R. Kalfat, M. M. Chehimi. Melamine-imprinted polymer grafts through surface photopolymerization initiated by aryl layers from diazonium salts. *Food Control* 31, 2013: 379–386.

342. Salmi Z., H. Benmehdi, A. Lamouri, P. Decorse, M. Jouini, Y. Yagci, M. M. Chehimi. Preparation of MIP grafts for quercetin by tandem aryl diazonium surface chemistry and photopolymerization. *Microchim. Acta* 180, 2013: 1411–1419.

343. Gam-Derouich S., M. Jouini, D. Ben Hassen-Chehimi, M. M. Chehimi. Aryl diazonium salt surface chemistry and graft photopolymerization for the preparation of molecularly imprinted polymer biomimetic sensor layers. *Electrochim. Acta* 73, 2012: 45–52.

344. Bakas, I., Z. Salmi, M. Jouini, F. Geneste, I. Mazerie, D. Floner, B. Carbonnier, Y. Yagci, and M. M. Chehimi. Picomolar detection of melamine using molecularly imprinted polymer-based electrochemical sensors prepared by UV-graft photopoly-merization. *Electroanalysis* 27, 2015: 429–439.

345. Gam-Derouich, S., S. Mahouche-Chergui, S. Truong, D. Ben Hassen-Chehimi, and M. M. Chehimi. *Polymer* 52, 2011: 4463–4470.

346. Jain R., D. C. Tiwari, and P. Karolia. Electrocatalytic detection and quantification of nitazoxanide based on graphene- polyaniline (Grp-PANI) nanocomposite sensor. *J. Electrochem. Soc.* 161, 2014: H839–H844.

347. Jlassi, K., A. Mekki, M. Benna-Zayani, A. Singh, D. K. Aswal, and M. M. Chehimi. Exfoliated clay/polyaniline nanocomposites through tandem diazonium cation exchange reactions and *in situ* oxidative polymerization of aniline. *RSC Adv.* 4, 2014: 65213–65222.

348. Lécayon, G., Y. Bouizem, C. Le Gressus, C. Reynaud, and C. Juret. Grafting and growing mechanisms of polymerised organic films onto metallic surfaces. *Chem. Phys. Lett.* 91, 1982: 506–510.

349. Palacin, S., C. Bureau, J. Charlier, G. Deniau, B. Mouanda, and P. Viel. Molecule-to-metal bonds: Electrografting polymers on conducting surfaces. *Chem. Phys. Chem.* 5, 2004: 1468–1481.

350. Deniau, G., L. Azoulay, L. Bougerolles, and S. Palacin. Surface electroinitiated emulsion polymerization: Grafted organic coatings from aqueous solutions. *Chem. Mater.* 18, 2006: 5421–5428.

351. Tessier, L., G. Deniau, B. Charleux, and S. Palacin. Surface electroinitiated emulsion polymerization (SEEP): A mechanistic approach. *Chem. Mater.* 21, 2009: 4261–4274.

352. Tessier, L., J. Chancolon, P.-J. Alet, A. Trenggono, M. Mayne-L'Hermite, G. Deniau, P. Jégou, and S. Palacin. Grafting organic polymer films on surfaces of carbon nanotubes by surface electroinitiated emulsion polymerization. *Phys. Stat. Sol. (a)* 205, 2008: 1412–1418.

353. Ghorbal, A., F. Grisotto, J. Charlier, S. Palacin, C. Goyer, and C. Demaille. Localized electrografting of vinylic monomers on a conducting substrate by means of an integrated electrochemical AFM probe. *Chem. Phys. Chem.* 10, 2009: 1053–1057.

354. Le, X. T., P. Viel, A. Sorin, P. Jegou, S. Palacin. Electrochemical behavior of polyacrylic acid coated gold electrodes: An application to remove heavy metal ions from wastewater. *Electrochim. Acta* 54, 2009: 6089–6093.

355. Le, X. T., P. Jégou, P. Viel, S. Palacin. Electro-switchable surfaces for heavy metal waste treatment: Study of polyacrylic acid films grafted on gold surfaces. *Electrochem. Commun.* 10, 2008: 699–703.

356. Le, X. T., P. Viel, P. Jégou, A. Sorin, S. Palacin. Electrochemical-switchable polymer film: An emerging technique for treatment of metallic ion aqueous waste. *Sep. Purif. Technol.* 69, 2009: 135–140.

357. Podvorica, F. I. Non-diazonium organic and organometallic coupling agents for surface modification, in: M. M. Chehimi, (Ed.), *Aryl Diazonium Salts. New Coupling Agents in Polymer and Surface Science.* Wiley-VCH, Weinheim, 2012.

358. Gallardo, I., J. Pinson, and N. Vilà. Spontaneous attachment of amines to carbon and metallic surfaces. *J. Phys. Chem. B* 110, 2006: 19521–19529.

359. Buttry, D. A., C. M. J. Peng, J.-B. Donnet, and S. Rebouillat. Immobilization of amines at carbon fiber surfaces. *Carbon* 37, 1999: 1929–1940.

360. Lee, L. and A. J. Downard. Preparation of ferrocene-terminated layers by direct reaction with glassy carbon: A comparison of methods. *J. Solid State Electrochem.* 18, 2014: 3369–3378.

361. Chrétien, J. M., M. A. Ghanem, P. N. Bartlett, and J. D. Kilburn. Covalent tethering of organic functionality to the surface of glassy carbon electrodes by using electrochemical and solid-phase synthesis methodologies. *Chem. Eur. J.* 14, 2008: 2548–2556.

362. Muthukumar P. and S. A. John. Synergistic effect of gold nanoparticles and amine functionalized cobalt porphyrin on electrochemical oxidation of hydrazine. *New J. Chem.* 38, 2014: 3473–3479.

363. Tanaka, M., T. Sawaguchi, Y. Sato, K. Yoshioka, and O. Niwa. Surface modification of GC and HOPG with diazonium, amine, azide, and olefin derivatives. *Langmuir* 27, 2011: 170–178.

364. Deinhammer, R. S., M. Ho, J. W. Anderegg, and M. D. Porter. Electrochemical oxidation of amine-containing compounds: A route to the surface modification of glassy carbon electrodes. *Langmuir* 10, 1994: 1306–1313.

365. Ghilane, J., F. Hauquier, and J.-C. Lacroix. Oxidative and stepwise grafting of dopamine inner-sphere redox couple onto electrode material: Electron transfer activation of dopamine. *Anal. Chem.* 85, 2013: 11593–1160.

366. Vanossi, D., R. Benassi, F. Parenti, F. Tassinari, R. Giovanardi, N. Florini, V. DeRenzi, G. Arnaud, and C. Fontanesi. Functionalization of glassy carbon surface by means of aliphatic and aromatic aminoacids. An experimental and theoretical integrated approach. *Electrochim. Acta* 75, 2012: 49–55.

367. Sarkar, C., C. Bora, and S. K. Dolui. Selective dye adsorption by pH modulation on amine-functionalized reduced graphene oxide – carbon nanotube hybrid. *Ind. Eng. Chem. Res.* 53, 2014: 16148–16155.

368. Cruickshank, K. and A. Downard. Electrochemical stability of citrate-capped gold nanoparticles electrostatically assembled on amine-modified glassy carbon. *Electrochim. Acta* 54, 2009: 5566–5570.

369. Cheng, L., J. Liu, and S. Dong. Layer-by-layer assembly of multilayer films consisting of silicotungstate and a cationic redox polymer on 4-aminobenzoic acid modified glassy carbon electrode and their electrocatalytic effects. *Anal. Chim. Acta* 417, 2000: 133–142.

370. Kumar, R. and D. Leech. Immobilisation of alkylamine-functionalised osmium redox complex on glassy carbon using electrochemical oxidation. *Electrochim. Acta* 140, 2014: 209–216.

371. Sun, B., H. Qi, F. Ma, Q. Gao, C. Zhang, and W. Miao. Double covalent coupling method for the fabrication of highly sensitive and reusable electrogenerated chemiluminescence sensors. *Anal. Chem.* 82, 2010: 5046–5052.

372. Geneste, F. and C. Moinet. Electrochemically linking TEMPO to carbon via amine bridges. *New J. Chem.* 29, 2005: 269–271.

373. Nasraoui, R., D. Floner, C. Paul-Roth, and F. Geneste. Flow electroanalytical system based on cyclam-modified graphite felt electrodes for lead detection. *J. Electroanal. Chem.* 638, 2010: 9–14.

374. Nasraoui, R., J.-F. Bergamini, S. Ababou-Girard, and F. Geneste. Sequential anodic oxidations of aliphatic amines in aqueous medium on pyrolyzed photoresist film surfaces for the covalent immobilization of cyclam derivatives. *J Solid State Electrochem.* 15, 2011: 139–146.

375. Downard, A. J., D. J. Garrett, and E. S. Q. Tan. Microscale patterning of organic films on carbon surfaces using electrochemistry and soft lithography. *Langmuir* 22, 2006: 10739–10746.

376. Garret, D., B. Flavel, J. Shapter, K. Baronian, and A. Downard. Robust forests of vertically aligned carbon nanotubes chemically assembled on carbon substrates. *Langmuir* 26, 2010: 1848–1854.

377. Mural, P. K. S., A. Banerjee, M. S. Rana, A. Shukla, B. Padmanabhan, S. Bhadra, G. Madras, and S. Bose. Polyolefin based antibacterial membranes derived from PE/PEO blends compatibilized with amine terminated graphene oxide and maleated PE. *J. Mater. Chem. A* 2, 2014: 17635–17648.

378. Ryu, S. H. and A. M. Shanmugharaj. Influence of hexamethylene diamine functionalized graphene oxide on the melt crystallization and properties of polypropylene nanocomposites. *Mater. Chem. Phys.* 146, 2014: 478–486.

379. Liu, H.-d., Z.-y. Liu, M.-b. Yang, and Q. He. Superhydrophobic polyurethane foam modified by graphene oxide. *J. Appl. Polym. Sci.* 130, 2013: 3530–3536.

380. Yang, L., J. Chen, X. Wei, B. Liu, and K. Kuang. Ethylene diamine-grafted carbon nanotubes: A promising catalyst support for methanol electro-oxidation. *Electrochim. Acta* 53, 2007: 777–784.

381. Tang, H., J. Chen, K. Cui, L. Nie, Y. Kuang, and S. Yao. Immobilization and electro-oxidation of calf thymus deoxyribonucleic acid at alkylamine modified carbon nanotube electrode and its interaction with promethazine hydrochloride. *J. Electroanal. Chem.* 587, 2006: 269–275.

382. Herlem, G., C. Goux, B. Fahys, A.-M. Gonçalves, C. Mathieu, E. Sutter, and J.-F. Penneau. Surface modification of platinum and gold electrode by anodic oxidation of pure ethylenediamine. *J. Electroanal. Chem.* 435, 1997: 259–265.

383. Lee, S. B., Y. Ju, Y. Kim, C. M. Koo, and J. Kim. Electrooxidative grafting of amine-terminated dendrimers encapsulating nanoparticles for spatially controlled surface functionalization of indium tin oxide. *Chem. Comm.* 49, 2013: 8913–8915.

384. Kim, Y. and J. Kim. Modification of indium tin oxide with dendrimer-encapsulated nanoparticles to provide enhanced stable electrochemiluminescence of Ru(bpy) 32+/

tripropylamine while preserving optical transparency of indium tin oxide for sensitive electrochemiluminescence-based analyses. *Anal. Chem.* 86, 2014: 1654–1660.

385. Herlem, G., K. Reybier, A. Trokourey, and B. Fahys. Electrochemical oxidation of ethylenediamine: New way to make polyethyleneimine-like coatings on metallic or semiconducting materials. *J. Electrochem. Soc.* 147, 2000: 597–601.

386. Adenier, A., M. M. Chehimi, I. Gallardo, J. Pinson, and N. Vilà. Electrochemical oxidation of aliphatic amines and their attachment to carbon and metal surfaces. *Langmuir* 20, 2004: 8243–8253.

387. Liu, T., Z. Zhao, W. W. Tjiu, J. Lv, and C. Wei. Preparation and characterization of epoxy nanocomposites containing surface-modified graphene oxide. *J. Appl. Polym. Sci.* 131, 2014: 40236.

388. Buriez, O., E. Labbé, P. Pigeon, G. Jaouen, and C. Amatore. Electrochemical attachment of a conjugated amino–ferrocifen complex onto carbon and metal surfaces. *J. Electroanal. Chem.* 169–175, 2008: 619–620.

389. Buriez, O., F. I. Podvorica, A. Galtayries, E. Labbé, S. Top, A. Vessières, G. Jaouen, C. Combellas, and C. Amatore. Surface grafting of a π-conjugated amino-ferrocifen drug. *J. Electroanal. Chem.* 699, 2013: 21–27.

390. Yang, G., Y. Shen, M. Wang, H. Chen, B. Liu, and S. Dong. Copper hexacyanoferrate multilayer films on glassy carbon electrode modified with 4-aminobenzoic acid in aqueous solution. *Talanta* 68, 2006: 741–747.

391. Li, X., Y. Wan, and C. Sun. Covalent modification of a glassy carbon surface by electrochemical oxidation of p-aminobenzene sulfonic acid in aqueous solution. *J. Electroanal. Chem.* 569, 2004: 79–87.

392. Yang, G., B. Liu, and S. Dong. Covalent modification of glassy carbon electrode during electrochemical oxidation process of 4-aminobenzylphosphonic in aqueous solution. *J. Electroanal. Chem.* 585, 2005: 301–305.

393. Herlem, M., B. Fahys, G. Herlem, B. Lakard, K. Reybier, A. Trokourey, T. Diaco, S Zairi, N. Jaffrezic-Renault. Surface modification of p-Si by a polyethylenimine coating: Influence of the surface pre-treatment. Application to a potentiometric transducer as pH sensor. *Electrochim. Acta* 47, 2002: 2597–2602.

394. Ge, F., R. C. Tenent, D. O. Wipf. Fabricating and imaging carbon-fiber immobilized enzyme ultramicroelectrodes with scanning electrochemical microscopy. *Anal. Sci.* 17, 2001: 27–35.

395. Liu, J., Cheng, L., Liu, B. and Dong, S. Covalent modification of a glassy carbon surface by 4-aminobenzoic acid and its application in fabrication of a polyoxometalates-consisting monolayer and multilayer films. *Langmuir* 16, 2000: 7471–7476.

396. Ghilane, J., P. Martin, H. Randriamahazaka, and J. C. Lacroix. Electrochemical oxidation of primary amine in ionic liquid media: Formation of organic layer attached to electrode surface. *Electrochem. Comm.* 12, 2010: 246–249.

397. Nowall, W. B., D. O.Wipf, and W. G. Kuhr. Localized avidin/biotin derivatization of glassy carbon electrodes using SECM. *Anal. Chem.* 70, 1998: 2601–2606.

398. Hayes, M. A. and W. G. Kuhr. Preservation of NADH voltammetry for enzyme-modified electrodes based on dehydrogenase. *Anal. Chem.* 71, 1999: 1720–1727.

399. Malmos, K., J. Iruthayaraj, S. U. Pedersen, and K. Daasbjerg. General approach for monolayer formation of covalently attached aryl groups through electrografting of arylhydrazines. *J. Am. Chem. Soc.* 131, 2009: 13926–13927.

400. Torbensen, K., K. Malmos, F. Kanoufi, C. Combellas, S. U. Pedersen, and K. Daasbjerg. Using time-resolved electrochemical patterning to gain fundamental insight into aryl-radical surface modification. *Chem. Phys. Chem.* 13, 2012: 3303–3307.

401. Malmos, K., J. Iruthayaraj, R. Ogaki, P. Kingshott, F. Besenbacher, S. U. Pedersen, and K. Daasbjerg. Grafting of thin organic films by electrooxidation of arylhydrazines. *J. Phys. Chem. C* 115, 2011: 13343–13352.

402. Maeda, H., K. Katayama, R. Matsui, Y. Yamauchi, and H. Ohmori. Surface improvement of glassy carbon electrode anodized in triethylene glycol and its application to electro-chemical HPLC analysis of protein-containing samples. *Anal. Sci.* 16, 2000: 293–298.

403. Ye, H. and R. M. Crooks. Electrocatalytic O_2 reduction at glassy carbon electrodes modified with dendrimer-encapsulated Pt nanoparticles. *J. Am. Chem. Soc.* 127, 2005: 4930–4934.

404. Verma, P., P. Maire, and P. Novák. Concatenation of electrochemical grafting with chemical or electrochemical modification for preparing electrodes with specific surface functionality. *Electrochim. Acta* 56, 2011: 3555–3561.

405. Andrieux, C. P., F. Gonzàlez, and J.-M. Savéant. Derivatization of carbon surfaces by anodic oxidation of arylacetates. Electrochemical manipulation of the grafted films. *J. Am. Chem. Soc.* 119, 1997: 4292–4934.

406. Brooksby, P. A., A. J. Downard, and S. S. C. Yu. Effect of applied potential on aryl-methyl films oxidatively grafted to carbon surfaces. *Langmuir* 21, 2005: 11304–11311.

407. Astudillo, P. D., A. Galano, and F. J. Gonzàlez. Radical grafting of carbon surfaces with alkylic groups by mediated oxidation of carboxylates. *J. Electroanal. Chem.* 610, 2007: 137–146.

408. Andrieux, C. P., F. Gonzàlez, and J.-M. Savéant. Homolytic and heterolytic radical cleavage in the Kolbe reaction. Electrochemical oxidation of arylmethyl carboxylate ions *J. Electroanal. Chem.* 498, 2001: 171–180.

409. Hilborn, J. W. and J. A. Pincock. Rates of decarboxylation of acyloxy radicals formed in the photocleavage of substituted 1-naphthylmethyl alkanoates. *J. Am. Chem. Soc.* 113, 1991: 2683–2686.

410. Hernández-Muñoz, L. S. and F. J.González. One-step modification of carbon surfaces with ferrocene groups through a self-mediated oxidation of ferroceneacetate ions. *Electrochem. Comm.* 13, 2011: 701–703.

411. Hernández-Muñoz, L. S., A. Galano, P. D. Astudillo-Sánchez, M. M. Abu-Omar, and F. J. González. The mechanism of mediated oxidation of carboxylates with ferrocene as redox catalyst in absence of grafting effects. An experimental and theoretical approach. *Electrochim. Acta* 136, 2014: 542–549.

412. Hernández-Muñoz, L. S., C. Frontana, and F. J. González. Covalent modification of carbon surfaces with cyclodextrins by mediated oxidation of β-cyclodextrin monoan-ions. *Electrochim. Acta* 138, 2014: 22–29.

413. Hernández-Morales D. M. and F. J. González. Covalent attachment of alkene and alkyne groups on carbon surfaces by electrochemical oxidation of unsaturated aliphatic carboxylates. *Electrochem. Comm.* 46, 2014: 48–51.

414. Hazimeh, H., S. Piogé, N. Pantoustier, C. Combellas, F. I. Podvorica, and F. Kanoufi. Radical chemistry from diazonium-terminated surfaces. *Chem. Mater.* 25, 2013: 605–612.

415. Lebègue, E., T. Brousse, J. Gaubicher, and C. Cougnon. Chemical functionalization of activated carbon through radical and diradical intermediates. *Electrochem. Comm.* 34, 2013: 14–17.

416. Kanazawa, A., T. Daisaku, T. Okajima, S. Uchiyama, S. Kawauchi, and T. Ohsaka. Characterization by electrochemical and X-ray photoelectron spectroscopic measurements and quantum chemical calculations of N-containing functional groups introduced onto glassy carbon electrode surfaces by electrooxidation of a carbamate salt in aqueous solutions. *Langmuir* 30, 2014: 5297–530.

417. Anne, A., B. Blanc, J. Moiroux, and J.-M. Savéant. Facile derivatization of glassy carbon surfaces by *N*-hydroxysuccinimide esters in view of attaching biomolecules. *Langmuir* 14, 1998: 2368–2371.

418. González-Fuentes, M. A., B. R. Díaz-Sánchez, A. Vela, and F. J. González. Radical grafting of carbon surfaces by oxidation of 5-nitroindole derived anions. *J. Electroanal. Chem.* 670, 2012: 30–35.

419. Sheridan, M. V., K. Lam, and W. E. Geiger. Covalent attachment of porphyrins and ferrocenes to electrode surfaces through direct anodic oxidation of terminal ethynyl groups. *Angew. Chem. Int. Ed.* 52, 2013: 1–5.

420. Fellah, S., F. Ozanam, J.-N. Chazalviel, J. Vigneron, A. Etcheberry, and M. Stchakovsky. Grafting and polymer formation on silicon from unsaturated grignards: I aromatic precursors. *J. Phys. Chem. B* 110, 2006: 1665–1672.

421. Fellah, S., A. Amiar, F. Ozanam, J.-N. Chazalviel, J. Vigneron, A. Etcheberry, and M. Stchakovsky. Grafting and polymer formation on silicon from unsaturated grignards: II. Aliphatic precursors. *J. Phys. Chem B* 111, 2007: 1310–1317.

422. Datsenko, S., N. Ignat'ev, P. Barthen, H.-J. Frohn, T. Scholten, T. Schroer, and D. Welting. Electrochemical reduction of pentafluorophenylxenonium, -diazonium, -iodonium, -bromonium, and -phosphonium salts. *Z. Anorg. Allg. Chem.* 624, 1998: 1669–1673.

423. Vase, K. H., A. H. Holm, K. Norrman, S. U. Pedersen, and K. Daasbjerg. Covalent grafting of glassy carbon electrodes with diaryliodonium salts: New aspects. *Langmuir* 23, 2007: 3786–3793.

424. Florini, N., M. Michelazzi, F. Parenti, A. Mucci, M. Sola, C. Baratti, V. De Renzi, K. Daasbjerg, S. U. Pedersen, and C. Fontanesi. Electrochemically assisted grafting of asymmetric alkynyl(aryl)iodonium salts on glassy carbon with focus on the alkynyl/aryl grafting ratio. *J. Electroanal. Chem.* 710, 2013: 41–47.

425. Weissmann, M., S. Baranton, and C. Coutanceau. Modification of carbon substrates by aryl and alkynyl iodonium salt reduction. *Langmuir* 26, 2010: 15002–15009.

426. Chan, C. K., T. E. Beechem, T. Ohta, M. T. Brumbach, D. R. Wheeler, and K. J. Stevenson. Electrochemically driven covalent functionalization of graphene from fluorinated aryl iodonium salts. *J. Phys. Chem. C* 117, 2013: 12038–1204.

427. Matrab, T., C. Combellas, and F. Kanoufi. Scanning electrochemical microscopy for the direct patterning of a gold surface with organic moities derived from iodonium salt. *Electrochem. Comm.* 10, 2008: 1230–1234.

428. Koczkur, K. M., E. M. Hamed, M. Chahma, D. F. Thomas, and A. Houmam. Electron transfer initiated formation of covalently bound organic layers on silicon surfaces. *J. Phys. Chem. C* 118, 2014: 20908–20915.

429. Chehimi, M. M., Hallais, G., Matrab, T., Pinson, J., Podvorica, F. I. Electro- and photografting of carbon or metal surfaces by alkyl groups. *J. Phys. Chem. C* 112, 2008: 18559–18565.

430. Hui, F., Noël, J.-M., Poizot, P., Hapiot, P., Simonet, J. Electrochemical immobilization of a benzylic film through the reduction of benzyl halide derivatives: Deposition onto highly ordered pyrolytic graphite. *Langmuir* 27, 2011: 5119–5125.

431. Jouikov, V., Simonet, J. Free propargyl radical: Facile cathodic generation and covalent linkage to solid surfaces. *Electrochem. Comm.* 15, 2012: 93–96.

432. Poizot, P., L. Laffont-Dantras, J. Simonet. The one-electron cleavage and reductive homo-coupling of alkyl bromides at silver–palladium cathodes. *J. Electroanal. Chem.* 624, 2008: 52–58.

433. Jouikov, V. and J. Simonet. Grafting of ω-alkyl ferrocene radicals to carbon surfaces by means of electrocatalysis with subnanomolar transition-metal (Pd, Pt, or Au) layers. *Chem. Plus. Chem.* 78, 2012: 70–76.

434. Rondinini, S., Mussini, P. R., Muttini, P., Sello, G. Silver as a powerful electrocatalyst for organic halide reduction: The critical role of molecular structure. *Electrochim. Acta* 46, 2001: 3245–3248.

435. Huang, Y.-F., D.-Y.Wu, A. Wang, B. Ren, S. Rondinini, Z.-Q. Tian, and C. Amatore. Bridging the gap between electrochemical and organometallic activation: Benzyl chloride reduction at silver cathodes. *J. Am. Chem. Soc.* 132, 2010: 17199–17210.

436. Andrieux, C. P., I. Gallardo, and J.-M. Savéant, Outer-sphere electron-transfer reduction of alkyl halides. A source of alkyl radicals or of carbanions? Reduction of alkyl radicals. *J. Am. Chem. Soc.* 111, 1989: 1620–1626.

437. Jouikov, V. and J. Simonet. Electrochemical conversion of glassy carbon into a poly-nucleophilic reactive material. Applications for carbon chemical functionalization. A mini-review. *Electrochem. Comm.* 45, 2014: 32–36.

438. Jouikov, V. and J. Simonet. Novel method for grafting alkyl chains onto glassy carbon. Application to the easy immobilization of ferrocene used as redox probe. *Langmuir* 28, 2012: 931–938.

439. Jouikov, V. and J. Simonet. Electrochemical cleavage of alkyl carbon-halogen bonds at carbon-metal and metal-carbon substrates: Catalysis and surface modification. *J. Electrochem. Soc.* 160, 2013: G3008–G3013.

440. Jouikov, V. and J. Simonet. Free allyl radical: Catalytic generation and subsequent immobilization onto Au, Pd, Pt, and carbon cathodes. *Electrochem. Comm.* 13, 2011: 1417–1419.

441. Wang, D. and J. M. Buriak. Trapping silicon surface-based radicals. *Langmuir* 22, 2006: 6214–6221.

442. Tsubota, T., S. Ida, O. Hirabayashi, S. Nagaoka, M. Nagata, and Y. Matsumoto. Chemical modification of diamond surface using a diacylperoxide as radical initiator and CN group-containing compounds for the introduction of the CN group. *Phys. Chem. Chem. Phys.* 4, 2002: 3881–3886.

443. Hetemi, D., F. Kanoufi, C. Combellas, J. Pinson, and F. I. Podvorica. Electrografting of alkyl films at low driving force by diverting the reactivity of aryl radicals derived from diazonium salts. *Langmuir* 30, 2014: 13907–13913.

444. Fabre, B. Ferrocene-terminated monolayers covalently bound to hydrogen-terminated silicon surfaces. Toward the development of charge storage and communication devices. *Acc. Chem. Res.* 43, 2010: 1509–1518.

445. Fabre, B, Y., Li, L. Scheres, S. P. Pujari, and H. Zuilhof. Light-activated electroactive molecule-based memory microcells confined on a silicon surface. *Angew. Chem. Int. Ed.* 52, 2013: 12024–12027.

446. Wang, X., E. C. Landris, R. Franking, and R. J. Hamers. Surface chemistry for stable and smart molecular and biomolecular interfaces via photochemical grafting of alkenes. *Acc. Chem. Res.* 43, 2010: 1205–1215.

447. Servinis, L., L. C. Henderson, T. R. Gengenbach, A. A. Kafi, M. G. Huson, and B. L. Fox. Surface functionalization of unsized carbon fiber using nitrenes derived from organic azides. *Carbon* 54, 2013: 378–388.

448. Amalric, J., C. Hammaecher, E. Goormaghtigh, and J. Marchand-Brynaert. Surface photografting of aryl azide derivatives on chalcogenide glasses. *J. Non-Cryst. Solids* 387, 2014: 148–154.

449 Devadoss, A. and C. E. D. Chidsey. Azide-modified graphitic surfaces for covalent attachment of alkyne-terminated molecules by "Click" Chemistry. *J. Am. Chem. Soc.* 129, 2007: 5370–5371.

450. Liu, H., S. Ryu, Z. Chen, M. L. Steigerwald, C. Nuckolls, and L. E. Brus. Photochemical reactivity of graphene. *J. Am. Chem. Soc.* 131, 2009: 17099–17101.

451. Zaba, T., A. Noworolska, C. M. Bowers, B. Breiten, G. M. Whitesides, and P. Cyganik. Formation of highly ordered self-assembled monolayers of alkynes on Au(111) substrate. *J. Am. Chem. Soc.* 136, 2014: 11918–11921.

452. Belmont, J. A., C. Bureau, M. M. Chehimi, S. Gam-Derouich, and J. Pinson. Patents and industrial applications of aryl diazonium salts and other coupling reagents, in: *Aryl Diazonium Salts. New Coupling Agents in Polymer and Surface Science.* M. M. Chehimi (Ed.), Wiley-VCH, Weinheim, 2012. Chapter 14.

Index

T - #0420 - 071024 - C248 - 234/156/11 - PB - 9780367377069 - Gloss Lamination